STANDARDIZATION
Mathematical Methods in
Assortment Determination

C. BONGERS
Erasmus University
Rotterdam

Springer-Science+Business Media, B.V

To Hans, Jaap and Marjo

Distributors for North America:
Martinus Nijhoff Publishing
Kluwer Boston, Inc.
160 Old Derby Street
Hingham, Massachusetts 02043

Distributors outside North America:
Kluwer Academic Publishers Group
Distribution Centre
P.O. Box 322
3300 AH Dordrecht, The Netherlands

ISBN 978-94-017-6405-6   ISBN 978-94-017-6403-2 (eBook)
DOI 10.1007/978-94-017-6403-2

Library of Congress Cataloging in Publication Data

Bongers, C
  Standardization.

  Bibliography: p.
  Includes index.
  1. Standardization —Mathematical models.
2. Economic lot size —Mathematical models. I. Title.
HD62.B58   658.5'62   79–25133
ISBN 0-89838-029-4

PREFACE

The purpose of this book is to present a number of mathematical models we have developed in the field of standardization, with examples and applications to show how these models can be used in practice.

Most books on standardization emphasize its advantages but neglect the fact that standardization restricts the consumer's freedom of choice. This restricted freedom of choice will usually imply losses and these losses, which are called adaptation losses, have to be set against the well-known benefits of standardization.

Taking the benefits as well as the adaptation losses into account when standardizing a product, has the effect that—for a specific class of standardization problems—economically optimal solutions can be found with respect to the choice of the number of sizes and the values of these sizes. In this book most attention has been paid to the choice of the sizes, given the number of sizes that has to be selected. An interesting result, in this respect, is that the choice of the sizes according to a preferred number series (or a Renard series) is in general not to be recommended from the economic point of view.

Although the book contains a rather large quantity of mathematical formulae, an extensive knowledge of mathematics and statistics is not required to understand the problems that will be considered. In fact, the first three chapters can be read by anyone who has a familiarity with elementary algebra. The other chapters require some knowledge of integration and differential calculus as far as the determination of maxima and minima.

On the statistical side, the book presupposes an introductionary course which covers among other things means, variances, conditional probabilities and continuous probability density functions such as the homogeneous distribution and the normal distribution.

I am indebted to many persons who have contributed to this book. My greatest personal debts are to Professor J. Sittig, Professor J.A. Hartog and my wife. In some excellent papers Professor J. Sittig has introduced the concept of adaptation loss in standardization problems and explained its practical importance. I have always been impressed by his creativity and I am grateful for the numerous useful ideas he has contributed.

Professor J.A. Hartog has stimulated my interest in statistics very much and taught me that statistics means more than a set of formulae only. He was always prepared to devote much time to review and discuss the manuscript. There is no doubt that this book has benefited greatly from his suggestions.

I am also indebted to my wife, Marjo, for her encouragement and support during the preparation of the book.

My colleagues at the Erasmus University, Rotterdam, have all contributed significantly to my understanding of statistics. Furthermore, I am grateful for the nice working climate they created in our department, making the writing of this book a pleasure.

My thanks also go to the department of computer science, A.I.V., of the Erasmus University for giving me time on their computer to perform the calculations on the numerical examples.

Finally, my thanks to Mrs. Marga Vermaat, for the remarkable accuracy with which she typed the manuscript.

Rotterdam, The Netherlands   Cornelis Bongers
June 1979

CONTENTS

1 INTRODUCTION

1.1 GENERAL BACKGROUND

The discipline of standardization has rapidly developed during the last decades. In the beginning of this century standardization has been recognized as a process capable of improving efficiency, communication and identification. Also, the importance of creating national and international standards began to be realized. Since then, countless product properties have been subject to standardization resulting in the situation that nowadays hardly any product can be found of which none of the properties is standardized.

In this study we intend to analyse some of the economic effects of standardization. Before discussing these effects we will give two descriptions of standardization. The first description comes from the International Organization for Standardization. It says:

> Standardization is the process of formulating and applying rules for an orderly approach to a specific activity for the benefit and with the cooperation of all concerned, and in particular for the promotion of optimum overall economy taking due account of functional conditions and safety requirements.
>
> It is based on the consolidated results of science, technique and experience. It determines not only the basis for the present but also for future development, and it should keep pace with progress.

1

Though this description covers a number of relevant concepts it is not concrete enough for our purpose. For instance, it does not mention the essence of standardization which consists of the reduction of the number of types of a product. Therefore we will use in this work the more formal description given by Sittig (1969):

Standardizing a product with respect to one or more specific properties implies,

1. selecting the properties which are to be standardized,
2. selecting the number of sizes,
3. selecting the sizes.

If the properties that are to be standardized are given — which will be assumed throughout the study — then standardization consists, according to this description, of the choice of a number of points — the sizes — in a one, two or multi dimensional property space. Sittig further points out that the number of sizes chosen and the pattern they form in the property space, determine the economic consequences of the standardization under consideration.

A few additional remarks about the description have to be made. In the first place, the word product must be interpreted in its broadest sense; a product is any good or service that might be taken into consideration for standardization. Thus this also includes services like education and jurisdiction.

Secondly, with respect to the properties of a product, we may distinguish between properties the values of which have a continuous character, such as the diameter of a saucepan, the maximum torque that an electromagnetic clutch can transmit or the thickness of window panes, and properties which can only assume discrete values such as the number of buttons for internal lines on a phone or the number of cigarettes in a packet. In the sequel we will consider nonnegatively valued product properties of the first kind only.

Finally, we call the set of all sizes the size pattern. The number of sizes within the size pattern will be denoted by the letter n.

1.2 THE ECONOMIC CONSEQUENCES OF STANDARDIZATION

The decision to standardize one or more properties of a product has in general important consequences [Verman, (1973)] ; in this section we will consider some of the economic effects.

Viewed from the angle of production, one of the most important effects of standardization, or more specifically variety reduction, is the increase of the (average) series length which opens the possibility to use large(r)-scale machinery. The use of large-scale instead of small-scale machinery will in many industries lead to a decrease of the manufacturing costs per unit. Some arguments on this matter are that large machines are cheaper per unit of capacity than small ones because their construction takes in general relatively less work and raw material. Furthermore, the running of large machines usually involves less labor effort than small machines per unit of capacity [Bruni, (1964)].

Apart from the savings mentioned above, savings may also occur if as a consequence of variety reduction fewer different production lines are required. This has the effect that fewer change-overs and set-ups are necessary which may have a considerable influence on the costs per unit [OEEC, (1954)].

Besides the manufacturing costs, the distribution costs will usually also decrease if the number of different types of a product is reduced. An advantageous effect, for example, is that less capital investment in stocks is required and that therefore the interest costs decrease. [Mey and Snel, (1964)].

Under the assumption that the quantity of sales remains the same when standardizing a product, it will be clear from the foregoing that the sum of the manufacturing and distribution costs—further to be denoted by the term production costs—of a product of which one or more properties are standardized will be less than the production costs of a similar nonstandardized product. Clearly, this is a favourable aspect of standardization.

Against the benefits due to standardization must however be mentioned its costs. In this respect we mean the costs of having to use standardized products rather than the costs of making and distributing the standards. These costs arise because of the fact that a consumer or user of a product which is standardized cannot in principle obtain the size he wants. This is a logical consequence of the assumed continuous character of the demand; it would be a mere coincidence if the size the consumer wants—further to be called the required size—equals one of the standardized sizes. If the consumer—although his required size is not available—still wishes to purchase the product, which will be assumed in the sequel, he has to choose one of the standardized sizes. In doing this he then suffers a loss which we call the adaptation loss. The following examples may clarify this point. Suppose that a number of identical (window) panes are needed for a block of newly built houses. The panes that will be ordered have to obey the requirements specified by the designer. The length and width of the panes for instance must be chosen such that they fit into the window frames and the thickness has to satisfy the appropriate safety rules. Let us assume that on the basis of these safety rules, a thickness of 5.2 mm is calculated. If the thickness of panes is standardized, then the calculated thickness will almost certainly not equal one of the standardized thicknesses. In that case, the designer has to order panes with a thickness larger than 5.2 mm, because the safety rules forbid him to install panes with a smaller thickness. Obviously, the designer then chooses the smallest standardized thickness larger than 5.2 mm. If this is a thickness of

5.5 mm, the adaptation loss, expressed in cm^3 glass, equals number of panes times surface of a pane (in cm^2) times (.55 - .52) cm.

As a second example, suppose that a designer needs an electro-motor of, say 75 kW, for a pump. If a motor of 75 kW does not belong to the size pattern the designer is forced to order a motor with a larger capacity than he needs since a motor with a smaller capacity would lead to overloading which reduces the length of life of the motor considerably. Let us assume that the nearest standardized size is a motor of 85 kW. If the designer purchases this motor then the amount of money he pays for the extra 10 kW of power is in the first instance his adaptation loss. However, there might also be some advantageous economic effects associated with the purchase of the 85 kW motor. It is for instance very well possible that—depending on the load of the motor—the power consumption of a 85 kW motor is less than that of a 75 kW motor. Furthermore the length of life of the motor might be increased because it would be underloaded. Clearly, these benefits also should be taken into account when calculating the adaptation loss.

From the last example we may conclude that not necessarily only losses occur if the required size differs from the purchased standardized size. Note, that this is also the case in the first example if we take into account the aspect of insulation. A larger thickness of panes namely leads to a (slightly) better insulation and therefore to a lower (present) value of the costs of heating and cooling. Nevertheless, the losses will always be larger than the benefits because otherwise the user of the product would have wanted to buy the standardized size, that is actually purchased, in the first place. The adaptation loss is thus always larger than or equal to zero.

An interesting point to note is that the adaptation loss—although it is specifically a loss that is suffered by the users of standardized products—corresponds in most cases partially or wholly with a part of the production cost. As an example we mention the standardization of the capacities of refrigerators. For this product it is also reasonable to assume that a consumer purchases a refrigerator with a standardized capacity larger than or equal to his required capacity. Let us further suppose that the adaptation loss that is suffered, if the required size and the purchased standardized size differ, consists of the cost price of the extra capacity and the increase of power consumption caused by the overcapacity during the lifetime of the refrigerator. The first part of the adaptation loss can be calculated by making use of the relationship between the cost price of a refrigerator and it capacity. Since the extra capacity the consumer purchases is actually produced, this part of the adaptation loss is a part of the production costs. On the other hand, the second part of his adaptation loss (i.e. the amount the consumer has to pay for the extra power consumption caused by the overcapacity) is not included in the production cost.

Another example of an adaptation loss that is not a part of the production costs is a loss which consists not in the first place of a loss of money. Such a loss occurs for instance when standardizing the length of shoes or the heights of sink-units. Standardization of such product properties will in general give rise to losses of comfort rather than losses of money.

As a measure of the adaptation losses suffered by the users of standardized products we will use the average or expected adaptation loss per unit of production. The choice of the unit of production depends on the property or properties that have to be standardized. In the sequel we will choose – when possible – the unit of production such that the adaptation loss per unit of production can be calculated as soon as the required values of the standardized properties are known. For instance, when standardizing the width and thickness of flat steel sections, the unit of production is a running meter (or cm) of flat steel, whereas when standardizing the thickness of window panes we take a pane of a m^2 as unit of production.

The expected adaptation losses – expressed as a percentage of the costs involved in buying and using the product – have been calculated for some existing size patterns of different products. The percentages varied between 2% and 20% with an average of about 10%.

These figures indicate that it certainly seems worthwhile to take the adaptation losses into account when standardizing a product. In doing this, it is not exceptional that the expected adaptation loss can be reduced by a third by re-arranging the sizes only. Unfortunately, however, losses of this type have yet been recognized by only a few writers on standardization of which we mention Bobeszko and Sittig. Except for a number of useful models constructed by these writers, little research has been done with respect to the description and calculation of the adaptation losses. More important, however, is the fact that in most practical standardization problems not much attention has been paid to the question of how these losses might be reduced or minimized.

1.3 PURPOSE AND SCOPE OF THE STUDY

When confronted with a standardization problem in which one has to choose the number and magnitude of the sizes, several objectives or rules may be used to govern this choice. Most attention will be paid to the objective consisting of the minimization of the average total costs or loss per unit of production over the period for which the standardization is meant to remain in force.

Since it does not make sense to include in this minimization process types of costs that depend neither on n nor on the sizes, we will incorporate in the average total costs or loss only that part of the adaptation loss and that part of the production costs that varies with n and/or the sizes.

In this study the essential elements of a class of standardization problems will be explained and these elements will be incorporated in several models which can be used to realize the objective mentioned above. The class of problems that will be considered is restricted to standardizations where,

1. the number of properties to be standardized equals one or two;
2. the sizes that have to be chosen are neither additive (i.e. a consumer cannot satisfy his demand by purchasing several standardized sizes, which together add up to his required size) nor (easily) divisible (i.e. a standardized size cannot be split up in one or more required sizes);
3. the choice of the sizes is not influenced by other (existing) size patterns.

In the rest of this section we will describe some of the elements that will be studied. We begin by supposing that the number of sizes to be chosen is known. The standardization problem then reduces to the choice of the sizes only. Since the costs that depend on the position of the sizes are all included in the expected adaptation loss, the minimization of this loss will—for any given n—lead to the same size pattern as the minimization of the total average cost per unit.

In order to minimize the expected adaptation loss we have to realize what elements have a major influence on this loss. Assuming one property has to be standardized, we may distinguish the following three elements:

1. the distribution of the demand,
2. the loss function,
3. the size pattern.

ad 1. The demand for a product property can often be reasonably well approximated by a probability density function, such as for instance the normal distribution. We call the density function which describes the demand of the consumers the "probability density function of the required sizes." Using this density function we may derive some useful figures, for example the sales percentage of each size for a given set of sizes.

ad 2. The loss function gives the adaptation loss, suffered by a consumer, as a function of the required size and the purchased standardized size. Depending on the characteristics of the property to be standardized, the loss function may have a simple or a rather complex form. An example of a simple loss function is the linear loss function which arises when the adaptation loss equals a cost factor times the difference between purchased standardized size and required size. Such a loss function for instance may occur if the property to be standardized is the width of book-shelves or the length of piles.

Finally, throughout the study, it is assumed that both the loss function and the density function do not change in time.

ad 3. The third element is the size pattern. As opposed to the density function and the loss function which are given data, the size pattern has to be chosen. There are various ways in which this can be done. Perhaps the most familiar method is to choose the size pattern according to a "preferred number series." In doing this the probability density function and the loss function are clearly not taken into consideration. This has the consequence that the adaptation losses of the consumers usually will be unnecessarily high. Obviously, a more economical method is to choose the size pattern such that, given the density function and the loss function, the expected value of the total adaptation loss suffered by the consumers over a certain period is minimal.

This study is primarily aimed at the analysis of the problems involved in the derivation and calculation of size patterns satisfying this objective. We will call such size patterns optimal size patterns.

Having considered the objective according to which the sizes will be chosen for a given n, we now will briefly discuss the choice of n. This will be done for the specific situation that the adaptation loss corresponds entirely with a part of the production cost. The average total cost per unit then consists of the average production cost per unit only. The choice of n influences the average production cost along various ways as can be seen from Fig. 1.1.

We observe that in the first place the expected adaptation loss depends via the size pattern on n. This will be clear if we, for the time being, neglect the influence n has via the cost parameters of the loss function on the expected adaptation loss and compare a standardization where one size is chosen with a standardization where an infinite number of sizes is (optimally) chosen. The expected adaptation loss in the latter case will be zero, whereas it reaches in the first case its maximal value.

As a less extreme case, we may analyze what happens if an extra size is added between two sizes of an optimal size pattern consisting of n sizes. The expected

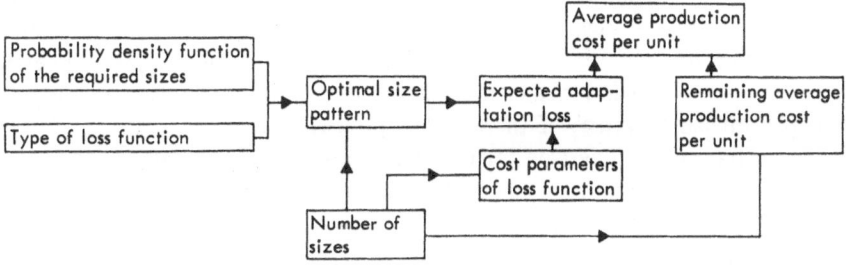

Figure 1.1. Diagram of the Relationship between n and the Average Production Cost per Unit (for a Specific Case)

adaptation loss then obviously decreases since some users of the product will purchase the newly added size instead of one of the sizes they bought before because they suffer in this way a smaller adaptation loss. However, if we take into consideration the influence n has via the cost parameters of the loss function on the expected adaptation loss, the situation becomes more complicated. Since the production cost will—as a consequence of diseconomies of scale—increase if an extra size is added to the size pattern, the cost parameters of the loss function may also become larger. This in turn will give rise to an increase of the expected adaptation loss. Taking both effects into account, the expected adaptation loss—expressed in monetary terms—might therefore increase, decrease or remain the same. In general, however, the expected adaptation loss will be a decreasing function of n. After the calculation of the expected adaptation loss for a given n, we have to add this amount to the remaining average production cost per unit in order to obtain the total average production cost per unit. By calculating this sum for a range of values of n, the optimal value of n (i.e. the n that minimizes the total average production cost per unit) can be found.

   The problem of choosing the optimal value of n will not be studied as intensively as the problem of finding the optimal size pattern. With respect to the determination of the optimal value of n, we will confine ourselves to the construction of a few simple models which, however, have a rather wide applicability in practical problems.

1.4   OUTLINE OF THE BOOK

Some different types of adaptation loss that may arise as a consequence of standardization are discussed in Chap. 2. Furthermore, a class of loss functions is introduced the members of which can be used to describe these adaptation losses. Special attention is paid to a number of loss functions which frequently arise in practice. Some of these loss functions can also be used in situations where the adaptation loss has—instead of an economical—a subjective character. In Chap. 3 a number of possible objectives and rules are given according to which a size pattern might be chosen. Also the objective we have adopted—the minimization of the expected adaptation loss—is treated in this chapter. Chapter 4 contains the formulae—further to be called the recursion formulae—an optimal size pattern has to satisfy if there is one property to be standardized. These recursion formulae are derived for the various loss functions treated in Chap. 2. Furthermore, some methods are given which can be used to find the optimal size pattern and the optimal value of n. In this chapter the probability density function, appearing in the recursion formulae, is left unspecified. In the Chaps. 5, 6, 7 and 8 we substitute successively the homogeneous -, the normal -, the log normal -, and the gamma distribution in the recursion formulae. The most simple among these distributions is the homogeneous distribution; it is therefore very well suited to illustrate the derivation and the position of optimal size patterns. The other three distributions merit their insertion from the fact that they often occur in practical situations. For each distribution some numerical examples are given to illustrate for a specific loss function the position of the optimal size pattern. Also, the method that is recommended to solve the recursion formulae is mentioned for each combination of density function and loss function. Chapter 9 contains a practical application, concerning the standardization of the power of electro motors, which illustrates the theory developed in the previous chapters.

Until now the probability density function of the required sizes was assumed to be known. In a practical situation this will generally not be the case; therefore we discuss in Chap. 10 what data might be used to determine the type of the density function and how to estimate from these data its parameters. In Chap. 11 the characteristics of an optimal size pattern are studied by means of a variance—and sensitivity analysis. Also the question of the uniqueness of optimal size patterns is shortly discussed. A comparison between size patterns chosen according to an arithmetical or geometrical series ("preferred numbers") and optimal size patterns is given in Chap. 12. It will appear that in most cases size patterns chosen according to a "preferred number" series are—from the economical point of view—not to be recommended. After the consideration of standardization problems where one property has to be standardized, we discuss in the second part of the book the standardization of two properties. In Chap. 13 the loss functions and the bivariate probability density functions are given which are to be used in the subsequent chapters. The problems involved when standardizing two properties turn out to be rather complicated if no restriction on the relative position of the sizes is imposed. We therefore discuss three restrictions, the grid -, the line - and the generalized grid restriction in the respective Chaps. 14, 15 and 16. In these chapters some numerical examples are given to illustrate the position of an optimal size pattern for a specific loss function and density function. Unrestricted selection of the sizes is considered in Chap. 17. A practical application concerning the standardization of the thickness and width of flat steel sections is given in Chap. 18. Some optimal size patterns for this product are calculated using the generalized grid restriction and the results are compared with the size pattern as given by the Euronorm. Finally, the mathematical results that were used to reduce the recursion formulae to a more simple form are inserted in the appendix. The appendix also contains a number of tables of optimal size patterns and a computer program with its description.

2 THE COSTS OF STANDARDIZATION

2.1 NOTATION

In Chap. 1, the concepts standardized size and required size were introduced. We define a standardized size as the value or the set of values that are attached to one or more properties of the product. According to this definition, the IQ level at which an education is aimed or the alloy of gold in a certain product may also be interpreted as sizes; the meaning of the word size in the definition is thus broader than the commonly used meaning. Mathematically, a standardized size is written as the vector $(x_i, y_i, \ldots)$. The various properties are indicated by the letters x, y etc. The subscript i gives the number of the size and the values of the variables x_i, y_i etc. equal the values attached to the respective properties of the product. If one property of the product is standardized, the brackets of the size vector will be omitted. Furthermore, the subscripts of the sizes are in this case chosen such that the sizes x_1, x_2, etc. form an increasing series. If two properties are standardized – which is the maximal number we consider – then one of these properties is chosen and the sizes are numbered in such a way that they form an increasing series with respect to this property. The required size (of a consumer or user) represents the required values of the product properties. The required size is written as the vector, $(x, y, \ldots)$ where the letters again indicate the product properties and the values of x, y, etc. are equal to the required values of the properties. Finally, in the sequel, we mean by a consumer – unless otherwise specified – someone who purchases (or wishes to purchase) and/or uses one unit of production. Hence, the required size must be interpreted as the required size of one unit of production demanded by a consumer. Furthermore, if no confusion can arise, the term "unit of production" will be abbreviated to "product" or "unit".

2.2 THE ADAPTATION LOSS

2.2.1 The Loss if the Consumer Is Served

Since the probability density function of the required sizes is assumed to be continuous, the required size of a consumer is in general not equal to one of the standardized sizes. This being the case, the consumer will have to decide whether or not to purchase the product and what size of the product if the first question is answered positively. When purchasing the product, the consumer will suffer a certain adaptation loss which depends on the purchased size as well as on the required size. Depending on the product property in question we may distinguish between several types of adaptation loss.

In the first place, it is possible that the purchased size has to be adapted to the required size by means of a certain adaptation operation. This for instance will be the case if a carpet (for close covering) is purchased with a (standardized) width that is not equal to the width of the room for which it is intended. The loss of material that occurs when cutting off a part of the carpet clearly forms a part of the adaptation loss. Another part of the adaptation loss consists of the costs of the adaptation operation itself. Note that these costs have to be taken into account only if they depend on the required size and/or the purchased standardized size because otherwise neither the decision as to the number of sizes to select nor the decision as to the choice of the sizes is influenced by them.

It is, however, not always possible or meaningful to adapt a standardized product. One can think for example of the number of watts of a lamp or the length of a bed. In both cases we may put the adaptation loss equal to the difference between the costs of the purchased standardized size and the costs of the required size if this required size would have been a size of the size pattern. The costs of the required size consist of the estimated price of the product and the estimated present value of its usage costs. An example of the latter is the present value of the cost of the current a lamp uses in its life. The costs of the standardized size consist – besides those two components – also of the additional costs caused by the use of the standardized size. We mention for instance the higher costs of ramming if piles are used with larger horizontal dimensions than required.

Finally, if the determination of the adaptation loss, as given above, gives not a good approximation to loss felt by the consumers, we can make use of a subjective loss function which also is a function of the purchased standardized size and the required size. Such a subjective loss function can for instance be used to describe the (subjective) loss felt by a consumer who buys a coat which does not fit perfectly. Also we can specify a subjective loss function on the loss of energy, time, etc. a highly intelligent child suffers in a school system which is based on the capacities of an average child.

2.2.2 The Loss if the Consumer Is Not Served

After the standardization of a product it may occur that a consumer cannot be served, i.e. there is no size of the size pattern acceptable to him. In general this will be the case when the product has to meet certain requirements which are not met by any size of the size pattern. These requirements often will be technically prescribed, such as the prescription that the sectional modulus in the main direction of a rolled steel beam must be equal to or larger than a given value.

It is assumed that the loss that arises if the consumer cannot be served equals a constant; this constant will be denoted by the letter B. We may interpret the loss of not being served as the difference between the costs of a special product, constructed according to the specifications of the consumer, and the estimated costs of the product if the required value of the product property in question would have been a size of the size pattern.

It is not always reasonable to introduce a loss of not being served over the whole range of demand. This is for instance the case if precision is the property to be standardized and a consumer requires a precision greater than the maximum technically attainable precision. In such a situation, it seems more obvious to neglect the demand for a precision greater than the maximum precision and to let the loss of not being served occur only if a consumer requires a precision between x_n and the maximum precision.

2.2.3 Shift Rules

In this investigation we shall assume that a consumer with a required size x, where $x_{i-1} < x \leqslant x_i$, purchases one of the standardized sizes x_{i-1} or x_i. We distinguish between the following two cases.

If the product has to meet certain requirements which are met by size x_i but not by size x_{i-1} the consumer with required size x ($x_{i-1} < x \leqslant x_i$) has to buy size x_i. In this case we say that the consumer applies shift rule O (One side). It also might occur that the consumer has to buy size x_{i-1} if his required size is x $(x_{i-1} \leqslant x < x_i)$ which is for instance the case if a product is required with a certain maximum allowed percentage pollution. We however do not discuss this situation here because it does not differ essentially from that mentioned above.

On the other hand, if the consumer is free to choose between x_{i-1} and x_i as for instance is the case when purchasing clothes or food, we speak of the application of shift rule T (Two sides). It is assumed in this case that the consumer chooses that size that minimizes his adaptation loss.

As will be obvious, shift rule T is more general than shift rule O. Mathematically, we can even restrict ourselves to the application of shift rule T, because the consumer can be forced to purchase size x_i by letting the incurred adaptation loss of purchasing size x_{i-1} approach infinity. Except in Sec. 2.4.1, where the general form of the loss function is introduced, both shift rules will, however, be discussed separately. This is due to the fact that numerical problems arise — when generating the size pattern — if the adaptation loss caused by buying size x_{i-1} $(x_{i-1} < x \leqslant x_i)$ approaches infinity.

From the foregoing it will be clear that, if the shift rule that is being applied is known and the size pattern is given, near a specific standardized size there must exist an area with the property that the consumer buys this size if his required size is a point in this area. We call such an area "the service area of the (standardized) size". If shift rule O is applied, the service areas have the shapes indicated by the arrows in Fig. 2.1.

The service area of size x_3 for instance is the interval (x_2,x_3) where the bound x_3 is included but the bound x_2 not. In case shift rule T is used, the boundaries of the service areas depend on the loss function that is used. However, in general the shapes of the service areas will be like those drawn in Fig. 2.2.

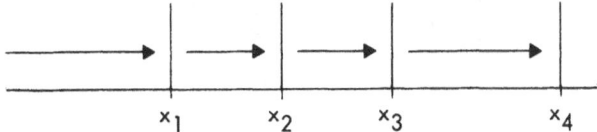

Figure 2.1. The Shapes of the Service Areas if Shift Rule O is Used

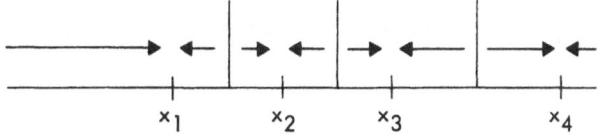

Figure 2.2. The Shapes of the Service Areas if Shift Rule T is Used

2.4 LOSS FUNCTIONS

2.4.1 The General Form of the Loss Function

As already mentioned, the adaptation loss is assumed to be a function of the required size, x, and the purchased standardized size x_i or x_{i-1}. We can write this function in its general form as,

$$L(x) = Min(V_1(x_{i-1},x), V_2(x_i,x)) \quad x_{i-1} < x \leqslant x_i, \quad i = 2, \ldots n,$$
$$L(x) = V_2(x_1,x) \qquad\qquad\qquad x \leqslant x_1 \qquad\qquad (2.1)$$

The function V_1 gives the adaptation loss if the consumer purchases size x_{i-1} (i.e. he purchases a size smaller than his required size, x). If the consumer purchases size x_i which is larger than or equal to his required size we use function V_2 to describe the adaptation loss. It is assumed that if the consumer needs a size smaller than x_1, which is the smallest size available, he will purchase x_1 with loss $V_2(x_1,x)$. On the other hand, if the required size is larger than x_n the adaptation loss is measured by one of the functions

$$L(x) = B \qquad \text{if the consumer does not purchase a}$$
$$\text{standardized size} \qquad\qquad\qquad (2.2)$$
$$L(x) = V_1(x_n,x) \quad \text{if the consumer purchases size } x_n. \qquad (2.3)$$

Which of these functions is used depends on the shift rule that is to be applied. We shall always use (2.2) in combination with shift rule O and (2.3) in combination with shift rule T.

The functions V_1 and V_2 are supposed to satisfy the following conditions,

$$V_1(t,x) > 0 \qquad \text{if } x > t,$$
$$V_2(t,x) > 0 \qquad \text{if } t > x,$$
$$V_1(t,x) = 0 \qquad \text{if } x = t,$$
$$V_2(t,x) = 0 \qquad \text{if } x = t,$$
$$V_1(t,x) > V_1(t,y) \quad \text{if } x > y > t,$$
$$V_2(t,x) > V_2(t,y) \quad \text{if } x < y < t, \qquad\qquad (2.4)$$

where t denotes the purchased standardized size and x and y represent required sizes.

The third and fourth conditions represent the fact that the adaptation loss equals zero if the required size equals the purchased standardized size. According to the fifth and sixth conditions the adaptation loss increases monotonously as a function of the difference between the required size and the purchased standardized size.

As will be obvious from (2.1) and (2.4) there exists a value, say x^*, between x_i and x_{i-1} for which the adaptation loss of purchasing size x_i equals that of size x_{i-1}. We have

$$V_1(x_{i-1}, x^*) = V_2(x_i, x^*) \quad x_{i-1} \leqslant x^* \leqslant x_i. \tag{2.5}$$

Solving (2.5) for x^* we find

$$x^* = z(x_i, x_{i-1}). \tag{2.6}$$

The function z is used to determine the bounds of the service areas of the respective sizes if shift rule T is applied.

If the loss function (2.1) is used, a number of conditions an optimal size pattern has to satisfy can be derived. This is done is Chap. 4.

In the further analysis we have restricted our attention to a specific class of loss functions. This is the class,

$$V_1(t,x) = k_1(x^\alpha - t^\alpha)^\beta \quad x \geqslant t \geqslant 0, \qquad \tag{2.7}$$
$$V_2(t,x) = k_2(t^\alpha - x^\alpha)^\beta \quad t \geqslant x \geqslant 0. \qquad \alpha, \beta \geqslant 0, \tag{2.8}$$

However, within this class a wide variety of loss functions is possible by varying α and β. Therefore only a few of them could be studied more intensively in this study. Selected were those loss functions which in our opinion are the most useful for practical applications. These are inserted in Table 2.1. The table also contains – for each loss function – the shift rule(s) that will be taken into consideration.

Table 2.1. The Loss Functions if One Product Property Is to Be Standardized

	$\alpha = 1$	$\alpha = 2$	α Not Specified
$\beta = 1$	linear loss function shift rules T and O	quadratic loss function shift rule O	nonlinear loss function shift rule O
$\beta = 2$	quadratic loss function shift rule T	–	–
β not spec.	nonlinear loss function shift rule T	–	–

In the sections to follow we shall shortly discuss these loss functions. Also, some examples will be given of possible fields of application.

2.4.2 The Linear Loss Function in Combination with Shift Rule O

If the adaptation loss is directly proportional to the difference between the purchased size and the required size, a linear loss function is indicated. When applying shift rule O, this loss function has the form,

$$L(x) = k(x_i - x) \quad x_{i-1} < x \leqslant x_i, \quad i = 1, .. n,$$
$$= B \qquad\qquad x_n < x, \qquad\qquad\qquad (2.9)$$

where k^1 is a cost factor equaling the loss per unit of difference between x_i and x. Note that x_0 must be interpreted as the value below which the demand for the product property is zero; x_0 is thus not a size. A linear loss function can for instance be used to describe the adaptation loss of a consumer who needs a steel beam of a certain length, say x. Clearly, the size that is purchased, say x_i, will have to be larger than or equal to the required size. If x_i is not equal to x, the purchased size must be adapted and if we assume that the part that is cut off is sold against scrap value because it is of no further use, the adaptation loss is directly proportional to the difference between x_i and x.

The loss function does not give a good approximation of the adaptation loss if the product in question can be easily divided into smaller products of the same kind. Suppose for instance that the standardized sizes of reinforcement steel for concrete equal 8 m and 14 m. A consumer who needs a bar of 6 1/2 m and a bar of 7 1/2 m will not purchase two standardized sizes of 8 m but the standardized size of 14 m and divide it into the required parts.

A similar situation occurs if the consumer needs more than two sizes. In that case he will try to choose one or more standardized sizes in such a way that these sizes are divisible into his required sizes with the least possible adaptation loss.

We have not yet succeeded in developing a model which handles this problem in a satisfactory manner. The consequence is that the models we consider give — in some cases — only an upper bound of the (expected) adaptation loss.

2.4.3 The Linear Loss Function in Combination with Shift Rule T

Application of shift rule T implies that the consumer purchases size x_i or x_{i-1} if his required size is x where $x_{i-1} < x \leqslant x_i$. We put the cost per unit of difference between x and x_{i-1} equal to k_1 and the cost per unit of difference between x and x_i equal to k_2.

---

[1] If shift rule O is to be applied, the subscript of k shall be omitted.

The consumer is indifferent between x_i and x_{i-1} if the adaptation loss is the same for both sizes. We have

$$k_1(x^* - x_{i-1}) = k_2(x_i - x^*),$$
$$z(x_i, x_{i-1}) = x^* = (k_2 x_i + k_1 x_{i-1})/(k_1 + k_2). \qquad (2.10)$$

The loss function is now specified as

$$\begin{aligned}
L(x) &= k_1(x - x_{i-1}) && \text{if } x_{i-1} < x \leqslant z(x_i, x_{i-1}), \\
&= k_2(x_i - x) && \text{if } z(x_i, x_{i-1}) < x \leqslant x_i, \\
L(x) &= k_1(x - x_n) && \text{if } x > x_n, \\
&= k_2(x_1 - x) && \text{if } x \leqslant x_1.
\end{aligned}$$
$$i = 2, \ldots n, \qquad (2.11)$$

As an example of a situation in which this loss function might be used we mention the purchasing of bread. Suppose the available standardized sizes are a half loaf and a whole loaf. If the required size equals three quarters of a loaf, the consumer must choose between the two standardized sizes. In case half a loaf is purchased the consumer will suffer a certain subjective loss because he can eat less bread than he wants. On the other hand, purchasing a whole loaf means eating old bread the next day, which also implies a subjective loss, or throw away one quarter of a loaf, which implies a financial loss. In the example given the sizes are additive; a consumer with required size one and a half loaf purchases size x_1 and size x_2 to fulfill his demand. We can take this additivity into account by considering only the fractional part of the required size. It is then assumed that a consumer with a required size larger than x_2 purchases a number, say q, of entire loafs such that $1 > x - qx_2 \geqslant 0$. Next he decides to purchase a whole loaf, half a loaf or no loaf to fulfill his remaining demand $x - qx_2$.

However, in general the specification of a loss function if the sizes are additive may give rise to the same kind of problems as we met in Sec. 2.4.2 especially if size x_i is no multiple of size x_{i-1} $(i = 2, \ldots, n)$. To avoid these complications we shall restrict ourselves in the sequel to the treatment of product (properties) which are not—or only at relatively high costs—divisible or additive.

2.4.4 The Nonlinear Loss Function in Combination with Shift Rule O

The parameter β of the loss function (2.8) is in this case equal to one so we have

$$\begin{aligned}
L(x) &= k(x_i^\alpha - x^\alpha) && x_{i-1} < x \leqslant x_i, \quad i = 1, \ldots, n, \\
&= B && x_n < x.
\end{aligned} \qquad (2.12)$$

The loss function might be used to describe the adaptation loss if the property to be standardized for instance is the strength of a cable, the capacity of a central heating unit or the reliability of electronic components.

Putting α equal to two we have the quadratic loss function. This loss function for instance occurs when purchasing a cylinder of steel if the diameter of such a cylinder is standardized and the height is fixed. If a consumer needs a cylinder with diameter x he will have to purchase a standardized size x_i larger than or equal to x. Assuming the costs of the cylinder to vary directly proportional to its volume it will be clear that (2.12) can be used to describe the adaptation loss, in this case consisting of the loss of material.

We can take the costs of reducing the standardized size to the required size into account by increasing the value of k—if these costs are proportional to the loss as given by (2.12)—or by adding a second term to the loss function. There are no particular difficulties in doing the latter but as the term to be added in principle differs from case to case we restricted ourselves to the treatment of loss functions consisting of one term only.

2.4.5 The Nonlinear Loss Function in Combination with Shift Rule T

The parameter α of the loss function discussed in this section equals unity so the function z is

$$z(x_i, x_{i-1}) = \frac{\sqrt[\beta]{k_2}\, x_i + \sqrt[\beta]{k_1}\, x_{i-1}}{\sqrt[\beta]{k_1} + \sqrt[\beta]{k_2}}. \tag{2.13}$$

The loss function now becomes

$$\begin{aligned}
L(x) &= k_1(x - x_{i-1})^\beta && \text{if } x_{i-1} < x \leqslant z(x_i, x_{i-1}), \\
&= k_2(x_i - x)^\beta && \text{if } z(x_i, x_{i-1}) < x \leqslant x_i, \quad i = 2, \ldots, n, \\
L(x) &= k_1(x - x_n)^\beta && \text{if } x > x_n, \\
&= k_2(x_1 - x)^\beta && \text{if } x \leqslant x_1.
\end{aligned} \tag{2.14}$$

A loss function of this form might be used to describe the adaptation loss that arises when purchasing a panty. For this product it is reasonable to assume that women with an equal surface below the waist can wear the same panty [Rozendal, (1974)]. If we take the surface of a panty as the property to be standardized and assume that small differences between required size and standardized size incur relatively less loss than large differences, which implies a nonlinear loss function, we may take (2.14) with $\beta > 1$ to describe the (subjective) adaptation loss of the consumer.

2.4.6 Other Loss Functions

In the preceding sections we mentioned a number of loss functions, of course without intending to give an exhaustive enumeration. For example, one of the loss functions that was not discussed is the "stepwise" loss function. The loss as given by this loss function is unequal to zero only if the purchased standardized size and the required size differ more than a given constant. Another possibility is to base a loss function upon the relative adaptation loss a consumer suffers. Assuming shift rule O to be applied this loss function has the form

$$L(x) = k(x_i - x)/x \quad x_{i-1} < x \leqslant x_i, \quad i = 1, \ldots, n,$$
$$L(x) = B \qquad\qquad x > x_n. \tag{2.15}$$

As will be clear this loss function represents the idea that a certain fixed difference between required size and standardized size becomes less important if the required size increases. The loss function thus does not give the 'real' loss a consumer suffers but merely a psychological loss.

We restricted ourselves to the treatment of the loss functions mentioned in the previous sections because we expect these loss functions frequently to occur in practice. Moreover the treatment of other loss functions seems superfluous because of the fact that the techniques to generate the sizes of the size pattern, when using other loss functions, will in general not give rise to particular difficulties. In most cases it is possible to use one of the techniques we shall describe in Chap. 4.

3 SEVERAL WAYS TO SELECT A SIZE PATTERN

3.1 INTRODUCTION

There are a number of different methods than can be used to select a size pattern. Some of these methods are based on an intuitive approach – for instance the method of choosing the sizes according to a geometrical series – while others are based on a clearly defined objective with an economical or social background. As already mentioned before, we have adopted the objective of the minimization of the expected adaptation loss; application of this objective leads to size patterns that are optimal in the sense that they minimize the costs for the society as a whole.

When comparing an optimal size pattern with a size pattern chosen according to another approach or objective, we may use the expected adaptation loss of the optimal size pattern as a criterion to judge the performance of the other size pattern. In Chap. 12 we will judge in this way the performances of size patterns chosen according to an arithmetical or geometrical series. On the other hand, also the performance of an optimal size pattern might be judged according to one or more criteria. Before discussing this matter in a later chapter, we will, however, first describe some of the approaches and objectives that may be used to choose the sizes.

3.2 ARITHMETICAL AND GEOMETRICAL SERIES

Arithmetical or geometrical series are an often used instrument when constructing a size pattern. If the sizes are chosen according to an arithmetical series the
difference between two successive sizes is a constant. This has the advantage that
the size pattern has a simple form and therefore is easy to remember. However,
the arithmetical series has lost a great deal of its popularity as many writers on
standardization nowadays are of the opinion that the difference between two
successive sizes ought to depend on the sizes themselves. This opinion is illustrated by an example given by Sanders (1972). He points out that a difference of
0.25 kW between a 0.75 kW and a 1.0 kW motor would be significant and reasonable whereas an equal difference between a 10 kW and a 10.25 kW motor
would be insignificant and useless. Sanders concludes that a range of sizes should
advance in a regular geometrical series rather than by arithmetical or any other
empirical progression. The geometrical series commonly used in standardization
are the "Renard series," also called the "Preferred numbers." These series repeat
themselves in successive powers of 10 and contain the number one. The four
series most often used are the R5, R10, R20 and R40 series which are listed in
Table 3.1. The common ratio of the R5 series is the fifth root of 10, the common ratio of the R10 series is the tenth root of 10 etc. The first term in Table
3.1 is for each series equal to one and the last term equals 10; it will, however,
be clear that the series can easily be adjusted so as to cover the domain of the
required sizes.

"The utility of the preferred numbers depends mainly on their general accep-
tance and use, in cases to which they are appropriate, in preference to arbitrary
values selected by individual designers" according to Sanders. Verman (1973)
considers the utility of the geometrical series from another point of view. He
states, "the practical use of preferred numbers has attracted a great deal of

Table 3.1. Series of Preferred Numbers

R5	R10	R20	R40	Exact Values to 4 Decimals
1.00	1.00	1.00	1.00	1.0000
			1.06	1.0593
		1.12	1.12	1.1220
			1.18	1.1885
	1.25	1.25	1.25	1.2589
			1.32	1.3335
		1.40	1.40	1.4125
			1.50	1.4962

Table 3.1. Continued

R5	R10	R20	R40	Exact Values to 4 Decimals
1.60	1.60	1.60	1.60	1.5849
			1.70	1.6788
		1.80	1.80	1.7783
			1.90	1.8836
	2.00	2.00	2.00	1.9953
			2.12	2.1135
		2.24	2.24	2.2387
			2.36	2.3714
2.50	2.50	2.50	2.50	2.5119
			2.65	2.6607
		2.80	2.80	2.8184
			3.00	2.9854
	3.15	3.15	3.15	3.1623
			3.35	3.3497
		3.55	3.55	3.5481
			3.75	3.7584
4.00	4.00	4.00	4.00	3.9811
			4.25	4.2170
		4.50	4.50	4.4668
			4.75	4.7315
	5.00	5.00	5.00	5.0119
			5.30	5.3088
		5.60	5.60	5.6234
			6.00	5.9566
6.30	6.30	6.30	6.30	6.3096
			6.70	6.6834
		7.10	7.10	7.0795
			7.50	7.4989
	8.00	8.00	8.00	7.9433
			8.50	8.4140
		9.00	9.00	8.9125
			9.50	9.4406
10.00	10.00	10.00	10.00	10.0000

attention of standards engineers and a considerable amount of literature has grown around the series in question but it still remains to be demonstrated that the solutions offered by their use in resolving standardization problems are the most economical". Verman further points out that a "strictly mathematical proof" is necessary.

It will be shown in a later chapter of this study that such a mathematical proof in favor of the geometrical series cannot be given, at least not under the assumptions we have adopted. It will even be made clear that in general the selection of the sizes according to a geometrical series is certainly not to be recommended from our point of view.

3.3 MINIMIZATION OF THE EXPECTED ADAPTATION LOSS

If we consider the choice of the sizes from an economic point of view, the size pattern ought to be determined, with any given n, in such a way that the expected adaptation loss is as small as possible. The average adaptation loss per unit can be easily calculated for a given period, say a month, if we assume the sizes required by the consumers to be known as well as the loss function and the size pattern. It will be clear that the average adaptation loss depends on the size pattern chosen. By varying the sizes it is possible to find a size pattern that minimizes this average adaptation loss. However, if we take another month and follow the same procedure it is very well possible that we arrive at another "optimal size pattern." The reason for this is that the required sizes—on basis of which such a size pattern is calculated—differ from month to month as they are only a sample from an underlying population of the required sizes. To arrive at an economically optimal solution we therefore first have to determine the characteristics of the population of the required sizes.

In this study it is assumed that this population has a probability density function of the continuous type. For the time being we suppose this density function to be known; in Chap. 10 attention will be paid to the problems that arise when fitting such a density function to an observed frequency distribution.

The size required by an arbitrary consumer now can be interpreted as a random variable drawn from the distribution of the required sizes. Consequently, the adaptation loss (per unit) of this consumer is also a random variable, the expectation of which can be calculated. Apart from some exceptions, our objective shall be to choose the size pattern in such a way as to minimize this expectation.

3.4 EQUAL TOTAL-OR EQUAL CONDITIONAL EXPECTED ADAPTATION LOSS PER SERVICE AREA

Although minimization of the expected adaptation loss is a useful objective on economic grounds, a drawback has to be mentioned. Specifically in some cases certain groups of consumers might be seriously injured when applying this objective. This will especially be the case for groups that are relatively small, which means that their contribution to the expected adaptation loss is also relatively small so that these groups practically do not influence the choice of the size pattern. For example, when standardizing clothes, the sizes might be selected rather unfavorably for very short or very tall persons.

When dividing the population in a number of subpopulations (groups) it seems reasonable to define the ith subpopulation as that part of the consumers whose required size is a point in the ith service area ($i = 1, \ldots, n$). For any given size pattern we then can calculate for every i,

1. the (total) expected adaptation loss of the ith subpopulation,
2. the conditional expected adaptation loss of the ith subpopulation.

By the term "conditional expected adaptation loss" we mean the expected adaptation loss per unit, given that the required size is a point in a certain service area. This loss is calculated on the basis of a conditional density function which can be derived from the probability density function of the population.

The sizes might now be chosen in such a way that the (total) expected adaptation losses in the subpopulations are equal to each other. When applying this rule the losses are equally divided over the different subpopulations. For instance if the length of a pair of trousers is the property to be standardized, it follows that the (total) expected adaptation loss of the subpopulation of consumers buying the nth size (i.e. the tall persons) will be equal to that of any other subpopulation. However, since there are fewer tall persons than medium sized persons, the adaptation loss of a tall person might on the average be much larger than that of a medium sized person. Therefore, a better approach is to choose the sizes such that the conditional expected adaptation losses of the subpopulations are equal to each other. In the example mentioned above, this would mean that the expected adaptation loss of a tall person buying a pair of trousers equals that of a medium sized person. When applying this objective, the adaptation losses are therefore fairly divided over the individual consumers.

The major disadvantage of the latter objective is that it in general does not lead to an economically optimal size pattern. Consequently, it may very well occur that the expected adaptation loss of a size pattern chosen according to this objective is much larger than that of an optimal size pattern. In that case it obviously is better to use the objective of the minimization of the expected adaptation loss and to smooth the loss differences in another way. A suggestion made by Hemelrijk is to put an extra tax charge to the price of some sizes and a subsidy to the price of others. The tax and subsidy amounts then might be chosen such that the conditional expected adaptation losses of the subpopulations are equal to each other. This method has the advantage that—from the economical point of view—the losses of standardization are minimal while at the same time a fair distribution of the losses takes place.

3.5 EQUAL SALES PER SIZE

Another possible method is to select the sizes in such a way that the sales per size are equal to each other. This method clearly does not take into account the adaptation losses suffered by the consumers. Sometimes, however, the determination of the adaptation loss (function) is very difficult or impossible. In that case the method might be used to compute the size pattern because it at least takes into account the characteristics of the distribution of the required sizes. This implies, loosely speaking, that the distance between two successive sizes is small near the top of the distribution and larger at the tails. In general, a size pattern obtained by minimizing the expected adaptation loss has similar characteristics although its expected adaptation loss obviously will be lower.

   As an application in which a slightly modified version of this criterion was used we mention the standardization of the horizontal dimensions of concrete columns [S.B.R., (1979)]. In principle, the most economical dimensions of a column can be calculated, given the forces it has to resist. Except for these forces there are, however, also aesthetic considerations which influence the dimensions of a column. As a consequence of the latter there existed no uniformly applied shift rule so that it became extremely difficult to calculate the expected adaptation loss. Instead of using the criterion of the expected adaptation loss it was therefore decided to base the choice of the size pattern on the (expected) sales per size.

4 THE CONSTRUCTION OF THE SIZE PATTERN AND THE DETERMINATION OF THE NUMBER OF SIZES WHEN STANDARDIZING ONE PROPERTY

4.1 INTRODUCTION

In this chapter we shall derive the conditions the size pattern has to satisfy in order to be optimal. In the literature about the subject these conditions are called the "recursion formulae." This term also shall be used by us; later in this chapter the meaning of the word "recursion" in this context will be explained.

We begin with the derivation of the recursion formulae for the general loss function (2.1). Next, the same is done for the loss functions defined by (2.7) and (2.8); this leads to a set of recursion formulae which obviously also is valid for the loss functions listed in Table 2.1. The results for these loss functions, obtained by substituting the relevant values of α and β into the recursion formulae, will be discussed separately. Also attention will be paid to the geometrical interpretation of some of the recursion formulae. The next section of this chapter contains some methods that can be used to find the optimal size pattern. Besides two numerical methods we will briefly discuss a graphical method which is—although its applicability is limited to only a few cases—attractive because of its simplicity. Finally in the last section, two models are given that may be used to find the optimal value of n.

4.2 NOTATION

Before proceeding further we shall briefly describe the notation used in this and the following chapters.

The symbols x_0 and x_{n+1} will be used in a number of formulae. We define x_0 as the smallest nonnegative value of x for which the probability density function of the required sizes, denoted by $h(x)$, is unequal to zero. Hence,

$$x_0 = \max \{0, \inf (x|h(x) > 0)\}. \qquad (4.1)$$

Similarly, x_{n+1} is defined as the largest value of x for which $h(x)$ is unequal to zero. Since we will only consider probability density functions for which x_{n+1} is positive we have

$$x_{n+1} = \sup (x|h(x) > 0). \qquad (4.2)$$

From (4.1) we see that x_0 is always chosen larger than or equal to zero which follows from the fact that we do not consider product properties that can assume negative values. For the majority of the probability density functions we use, the probability of a required size being negative equals zero which automatically leads to a nonnegative x_0. An exception is the normal distribution which is defined on the interval $(-\infty, \infty)$. When using this distribution we therefore will assume that the probability of a required size being negative is that small that the tail of the distribution above the negative part of the x-axis can be neglected.

The symbol S_i is used to indicate the ith service area and the probability that the required size is a point in S_i is denoted by p_i. Therefore

$$p_i = \int_{S_i} h(x)dx. \qquad (4.3)$$

If shift rule T is to be applied we may write,

$$p_i = \int_{z(x_i,x_{i-1})}^{z(x_{i+1},x_i)} h(x)dx \qquad i = 2,3,\ldots,n-1. \qquad (4.4)$$

It will be clear that this formula is not valid for $i = 1$. This is because of the fact that the service area of size x_1 equals the interval $(x_0, z(x_2,x_1))$ and not the interval $(z(x_1,x_0), z(x_2,x_1))$ as would be indicated by (4.4). A similar problem arises if $i = n$.

We have therefore defined the function z_i as,

$$z_i = z(x_i, x_{i-1}) \qquad i = 2,\ldots,n,$$
$$z_1 = x_0,$$
$$z_{n+1} = x_{n+1}, \qquad (4.5)$$

so that now

$$p_i = \int_{z_i}^{z_{i+1}} h(x)dx \tag{4.6}$$

for all values of i. The p_i defined by (4.6) are valid if shift rule T is used. For shift rule O we have,

$$p_i = \int_{x_{i-1}}^{x_i} h(x)dx \quad i = 1, \ldots, n. \tag{4.7}$$

The sum

$$\sum_{i=1}^{n} p_i \tag{4.8}$$

then equals the probability that a consumer is served. We call this probability the coverage of the size pattern. The probability that a consumer is not served is denoted by p_{n+1}, thus

$$p_{n+1} = 1 - \sum_{i=1}^{n} p_i. \tag{4.9}$$

We further shall use lower case symbols to denote probability density functions and capitals to denote cumulative distribution functions.

Finally, we write the expected adaptation loss, using the expectation operator E, as E(L).

4.3 THE RECURSION FORMULAE FOR THE LOSS FUNCTION DEFINED BY (2.1)

Assuming the loss function to be of the form (2.1) implies that the expected adaptation loss can be written as,

$$E(L) = \sum_{i=1}^{n} \int_{z_i}^{x_i} V_2(x_i,x)\, h(x)dx$$

$$+ \sum_{i=1}^{n-1} \int_{x_i}^{z_{i+1}} V_1(x_i,x)\, h(x)dx + \int_{x_n}^{x_{n+1}} L(x)\, h(x)dx \tag{4.10}$$

where $L(x)$ is defined by (2.2) or (2.3). The formula is, besides for shift rule T, also valid for shift rule O; by letting k_1 approach infinity, z_i namely approaches x_{i-1} for $i = 2, \ldots, n$.

The set of sizes which minimizes $E(L)$ can now easily be found by partial differentation of (4.10) with respect to $x_1, \ldots, x_n$. To obtain the derivative with respect to x_i ($i \neq n$) we first collect the terms in which x_i appears. These are

$$\int_{x_{i-1}}^{z_i} V_1(x_{i-1},x)\, h(x)dx + \int_{z_i}^{x_i} V_2(x_i,x)\, h(x)dx$$

$$+ \int_{x_i}^{z_{i+1}} V_1(x_i,x)\, h(x)dx + \int_{z_{i+1}}^{x_{i+1}} V_2(x_{i+1},x)\, h(x)dx. \tag{4.11}$$

Differentiating[1] we have after some rearranging

$$\frac{\partial E(L)}{\partial x_i} = \int_{z_i}^{x_i} \frac{\partial}{\partial x_i} V_2(x_i,x)\, h(x)dx + \int_{x_i}^{z_{i+1}} \frac{\partial}{\partial x_i} V_1(x_i,x)\, h(x)dx$$

$$i = 1, \ldots, n-1. \tag{4.12}$$

The derivative with respect to x_n depends on the shift rule that is used. This leads to,

$$\frac{\partial E(L)}{\partial x_n} = \int_{z_n}^{x_n} \frac{\partial}{\partial x_n} V_2(x_n,x)\, h(x)dx + \int_{x_n}^{x_{n+1}} \frac{\partial}{\partial x_n} V_1(x_n,x)\, h(x)dx \quad \text{for shift rule T,}$$

$$\frac{\partial E(L)}{\partial x_n} = \int_{z_n}^{x_n} \frac{\partial}{\partial x_n} V_2(x_n,x)\, h(x)dx - Bh(x_n) \qquad \text{for shift rule O.} \tag{4.13}$$

The formulae (4.12) and (4.13) form the starting point of the analysis. Putting these expressions equal to zero we have the necessary conditions the size pattern must satisfy in order to be optimal. As will be clear from (4.12) and (4.13) each condition constitutes a relation between three successive sizes. If x_0 and size x_1 are known we can find size x_2 using the derivative of $E(L)$ with respect to x_1. Using the derivative with respect to x_2 we find size x_3 by substituting into this condition the values of x_1 and x_2, etc. The nature of this process is recursive which explains the term "recursion formulae" we use to denote these conditions.

---

[1] To avoid unnecessary complications it is assumed that the first and second derivatives of the functions V with respect to the sizes exist everywhere in the interval (x_0, x_{n+1}).

4.4 THE RECURSION FORMULAE FOR THE LOSS FUNCTION DEFINED BY (2.7) AND (2.8)

Assuming V_1 to be of the form (2.7) and V_2 of the form (2.8) we have

$$\frac{\partial V_1(x_i,x)}{\partial x_i} = -k_1 \alpha \beta x_i^{\alpha-1} (x^\alpha - x_i^\alpha)^{\beta-1} ,$$

$$i = 1, \ldots, n,$$

$$\frac{\partial V_2(x_i,x)}{\partial x_i} = k_2 \alpha \beta x_i^{\alpha-1} (x_i^\alpha - x^\alpha)^{\beta-1} .$$

$$(4.14)$$

Substituting (4.14) into (4.12) and putting the result equal to zero we obtain

$$k_2 \int_{z_i}^{x_i} (x_i^\alpha - x^\alpha)^{\beta-1} h(x)dx = k_1 \int_{x_i}^{z_{i+1}} (x^\alpha - x_i^\alpha)^{\beta-1} h(x)dx.$$

$$(4.15)$$

This formula is valid for all values of i if shift rule T is applied. When using shift rule O we have to replace the nth equation by,

$$k_2 \alpha \beta x_n^{\alpha-1} \int_{x_{n-1}}^{x_n} (x_n^\alpha - x^\alpha)^{\beta-1} h(x)dx = B\, h(x_n) \qquad (4.16)$$

which follows directly from (4.13). Note that both terms of the equation (4.15) were divided by $\alpha \beta x_i^{\alpha-1}$. Implicitly it is thus assumed that $x_i \neq 0$ (i = 1, ..., n) if $\alpha \neq 1$. The function z can be calculated by equating (2.7) and (2.8). This gives

$$z_i = \left[\frac{\sqrt[\beta]{k_2}\, x_i^\alpha + \sqrt[\beta]{k_1}\, x_{i-1}^\alpha}{\sqrt[\beta]{k_1} + \sqrt[\beta]{k_2}} \right]^{\frac{1}{\alpha}} \qquad i = 2, \ldots, n. \qquad (4.17)$$

Next we will consider the necessary conditions for the various loss functions listed in Table 2.1. Also some methods that might be used to compute the optimal size pattern will be given.

4.5 THE RECURSION FORMULAE FOR SOME SPECIFIC LOSS FUNCTIONS AND SHIFT RULES

4.5.1 The Linear Loss Function in Combination with Shift Rule O

To obtain the linear loss function the parameters α and β are both put equal to one. Since shift rule O is applied we let the adaptation loss of purchasing a standardized size smaller than the required size approach infinity. As already mentioned before, this is done by letting the parameter k_1 approach infinity so that, according to (4.17), the value of z_i approaches x_{i-1} ($i = 2, \ldots, n$). The left-hand side of (4.15) then has the form,

$$k_2 \int_{x_{i-1}}^{x_i} h(x)dx. \tag{4.18}$$

Because the difference between z_{i+1} and x_i is small as a consequence of the large value of k_1, the right-hand side of (4.15) can be approximated by,

$$k_1(z_{i+1} - x_i) h((z_{i+1} + x_i)/2) \tag{4.19}$$

which can be rewritten as

$$\frac{k_1 k_2 (x_{i+1} - x_i)}{k_1 + k_2} h \left(\frac{k_2 (x_{i+1} + x_i) + 2k_1 x_i}{2(k_1 + k_2)} \right). \tag{4.20}$$

If k_1 approaches infinity (4.20) becomes

$$k_2(x_{i+1} - x_i) h(x_i). \tag{4.21}$$

The recursion formulae[2] are therefore,

$$\int_{x_{i-1}}^{x_i} h(x)dx = (x_{i+1} - x_i) h(x_i) \quad i = 1, \ldots, n-1,$$

$$k \int_{x_{n-1}}^{x_n} h(x)dx = B h(x_n) \tag{4.22}$$

where $k = k_2$.

According to (4.22) the position of the optimal size pattern depends on the value of the factor B/k and on the form of the density function. In general the sizes are uniquely determined by (4.22) and as x_{i+1} is a single valued function of x_i and x_{i-1} we have

$$x_{i+1} = x_i + p_i/h(x_i) \quad i = 1, \ldots, n-1. \tag{4.23}$$

This fact enables us, as we will see in Sec. 4.6, to determine the optimal size pattern with a simple iterative method.

An interesting point to note is that the ith equation of (4.22) ($i \neq n$) can be interpreted geometrically. The left-hand side namely equals the surface of the probability density function between x_{i-1} and x_i and the right-hand side equals the surface of a rectangle with base $x_{i+1} - x_i$ and height $h(x_i)$. Size x_i is optimally located, given size x_{i-1} and size x_{i+1}, if both surfaces equal each other.

---

[2] Sometimes there exists no solution of the nth equation of (4.22) for x_n. In Chap. 5 we will consider this problem.

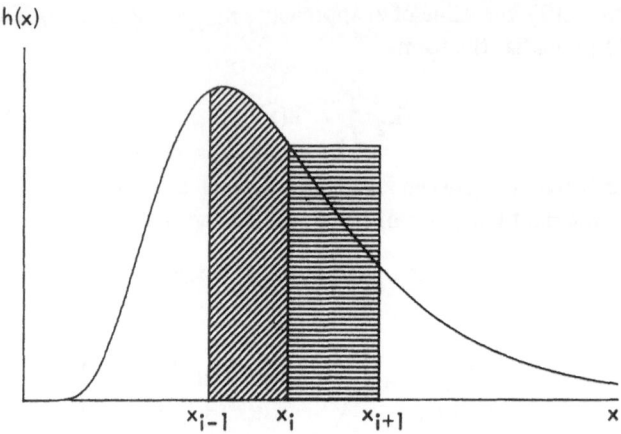

Figure 4.1. The Optimal Position of x_i, Given x_{i-1} and x_{i+1} (Linear Loss Function, Shift Rule O)

This situation is displayed in Fig. 4.1. Finally, to check whether the conditions (4.22) lead to a minimum of the expected adaptation loss, we calculate the matrix of second derivatives, A (see also 4.34),

$$A = \begin{bmatrix} a_{11} \; a_{12} \cdots a_{1n} \\ a_{21} \qquad\qquad\cdot \\ \cdot \qquad\qquad\quad\cdot \\ \cdot \qquad\qquad\quad\cdot \\ \cdot \qquad\qquad\quad\cdot \\ a_{n1} \qquad \cdots a_{nn} \end{bmatrix} \qquad (4.24)$$

where

$$a_{ii} = k(2h(x_i) - (x_{i+1} - x_i) h'(x_i)) \quad i = 1, \ldots, n-1,$$
$$a_{nn} = k\, h(x_n) - B\, h'(x_n),$$
$$a_{i,i+1} = a_{i+1,i} = -k\, h(x_i) \qquad\qquad i = 1, \ldots, n-1,$$

and all other elements equal zero. As can be seen, the derivative of $h(x)$ with respect to x, evaluated at x_i, is denoted by $h'(x_i)$.

The matrix A is positive definite if the stationary point is a minimum. If the solution of the recursion formulae is unique this will in general be the case. Sometimes, however, there are more solutions of the recursion formulae. The expected adaptation loss then might reach a saddle point or a maximum at one of these solutions in which case the matrix A is not positive definite. We will give some examples of size patterns at which the expected adaptation loss reaches a maximum in Chap. 11.

4.5.2   The Linear Loss Function in Combination with Shift Rule T

According to (4.15) we have

$$k_2 \int_{z_i}^{x_i} h(x)dx = k_1 \int_{x_i}^{z_{i+1}} h(x)dx \quad i = 1, \ldots, n. \tag{4.25}$$

From (4.25) we see that x_{i+1} can be written as a function of x_i and x_{i-1}. Denoting the cumulative distribution function of the required sizes by H we obtain

$$x_{i+1} = \left((k_1 + k_2) H^{-1} \left(\frac{k_2}{k_1} (H(x_i) - H(z_i)) + H(x_i) \right) - k_1 x_i \right) / k_2 \quad i = 1, \ldots, n. \tag{4.26}$$

When computing the size pattern with the simple iterative method mentioned in the previous section we need—as can be seen from (4.26)—an analytic expression or a numerical approximation of the inverse cumulative distribution function of the required sizes. If both are not available the optimal size pattern can be determined with the method of Newton which is described in Sec. 4.6.3.

The recursion formulae (4.25) might—after a slight modification—also be interpreted geometrically. Putting namely k_1 equal to k_2, each size has to be chosen in such a way that it splits up its service area into two parts with an equal probability. Hence, given size x_{i-1} and x_{i+1}, size x_i must be chosen such that the two shaded areas in Fig. 4.2 have the same surface.

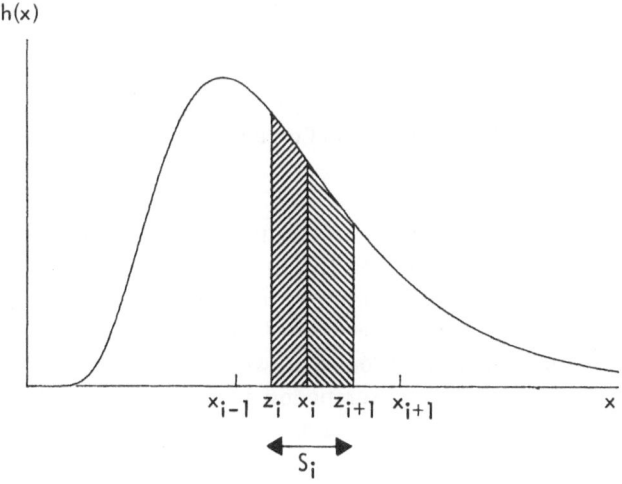

Figure 4.2. The Optimal Position of x_i, Given x_{i-1} and x_{i+1} (Linear Loss Function, Shift Rule T)

4.5.3 The Nonlinear Loss Function in Combination with Shift Rule O

The parameter β in (2.7) and (2.8) now equals one whereas α has an arbitrary positive value. We may find the recursion formulae by substituting $\beta = 1$ into (4.15) and letting k_1 approach infinity. In this case it is, however, easier to derive the recursion formulae directly from the formula of the expected adaptation loss.

With $\beta = 1$ we have

$$E(L) = k \sum_{i=1}^{n} \left\{ \int_{x_{i-1}}^{x_i} (x_i^{\alpha} - x^{\alpha})h(x)\,dx \right\} + B \int_{x_n}^{x_{n+1}} h(x)\,dx. \tag{4.27}$$

The partial derivatives are

$$\frac{\partial E(L)}{\partial x_i} = k(\alpha x_i^{\alpha-1} p_i - (x_{i+1}^{\alpha} - x_i^{\alpha})h(x_i)) \quad i = 1,\ldots,n-1,$$

$$\frac{\partial E(L)}{\partial x_n} = k\alpha x_n^{\alpha-1} p_n - Bh(x_n). \tag{4.28}$$

Putting the first n - 1 derivatives of (4.28) equals to zero we obtain

$$x_{i+1} = x_i \{\alpha p_i/x_i h(x_i) + 1\}^{1/\alpha} \quad i = 1,\ldots,n-1. \tag{4.29}$$

If $\alpha = 1$ then (4.29) equals (4.23). The recursion formulae for the generation of the size pattern if the loss function is quadratic are obtained by putting α equal to two.

4.5.4 The Nonlinear Loss Function in Combination with Shift Rule T

The loss function has the form (2.14) so that the recursion formulae are obtained by substituting $\alpha = 1$ into (4.15). In general, however, (4.15) cannot be reduced to a simple form for an arbitrary value of β. Depending on the probability density function of the required sizes, we shall restrict ourselves therefore in most cases to values of β equal to one or two.

Putting β equal to two the quadratic loss function arises. In the special case that $k = k_1 = k_2$ the recursion formulae are

$$\int_{z_i}^{x_i} (x_i - x)h(x)\,dx = \int_{x_i}^{z_{i+1}} (x - x_i) h(x)\,dx \quad i = 1,\ldots,n. \tag{4.30}$$

These may be rewritten as

$$\int_{z_i}^{z_{i+1}} (x_i - x) h(x) dx = 0 \quad i = 1, \ldots, n$$

or

$$x_i = \left(\int_{z_i}^{z_{i+1}} x \, h(x) dx \right) / p_i \quad i = 1, \ldots, n. \tag{4.31}$$

Given the sizes x_{i-1} and x_{i+1}, size x_i has thus to equal the expectation of the required size given that the required size is a point in S_i.

4.6 TWO METHODS TO COMPUTE THE OPTIMAL SIZE PATTERN

4.6.1 Introduction

The two methods which will be discussed in the following sections are the method of interval bisection and Newton's method. In principle, Newton's method always can be used to locate the optimal size pattern for the class of problems we consider. The method of interval bisection, on the other hand, is in principle to be used only if x_{i+1} can be written explicitly as a function of x_i and x_{i-1}. If this function has a simple form the method of interval bisection generally will require less c.p.u. time to find the optimal size pattern than Newton's method. For more complicated functions, however, the methods are about equivalent.

4.6.2 Interval Bisection

To illustrate this method we shall assume that the linear loss function is used in combination with shift rule O.

To compute the size pattern we first select a starting value for the smallest size of the pattern, say x_1^1. Next, the remaining sizes of the size pattern can be calculated using (4.23) which implies that the first $n - 1$ recursion formulae of (4.22) are satisfied. If x_1^1 is chosen optimally the derivative of $E(L)$ with respect to x_n, evaluated at x_n^1,

$$k \int_{x_{n-1}^1}^{x_n^1} h(x) dx - Bh(x_n^1) \tag{4.32}$$

equals zero and the optimal size pattern is found. If (4.32) does not equal zero we choose a new starting value x_1^2. Supposing the sign of (4.32) to be positive we may try to reduce the value of (4.32) by choosing x_1^2 (slightly) smaller than x_1^1. This generally has the effect that the distances between the sizes decrease and that the probabilities of the service areas become smaller. The first term of (4.32) therefore decreases. Furthermore, the smaller distances between the sizes imply that x_n^2 is smaller than x_n^1. If we assume that h(x) is a decreasing function of x in the neighborhood of x_n^1 the second term of (4.32) becomes larger. Taking both effects into account, the value of (4.32) evaluated at x_n^2 will therefore in general be closer to zero than the value found at x_n^1. If, on the other hand, the value of (4.32) computed with x_n^1 is negative, we follow the reverse procedure and choose x_1^2 larger than x_1^1.

It will be clear that the method will not always work successfully. If, for instance, h(x) is increasing in the neighborhood of x_n^1 it may occur that the direction of the step is wrong. For the density functions we will consider in the next 4 chapters, however, the method works satisfactorily provided that the factor B/k is not too small.

After the determination of the direction of the step, we have to choose an initial step size, say t. This step size is divided by two after each iteration. To be sure that the optimal size pattern can be found, t must therefore be chosen that large that the interval $x_1^1 \pm 2t$ contains the optimal x_1 with certainty.

It may happen of course, that at a certain iteration x_n^j becomes larger than x_{n+1}. In that case we select the new x_1^{j+1} as x_1^j minus the current value of the step, $t(1/2)^{j-1}$. If x_1^j becomes smaller than x_0 we choose x_1^{j+1} as x_1^j plus $t(1/2)^{j-1}$.

For the stop criterion we may use the absolute value of the relative difference between two successive starting sizes, x_1^j and x_1^{j+1}. As soon as this value becomes smaller than a given constant the iterations are terminated. When applying this stop rule it is recommended to calculate and print the expected adaptation loss at each iteration to check if the constant is chosen sufficiently small.

4.6.3 Newton's Method

Another method that can be used to compute the optimal size pattern is the method of Newton. Applying this method we use the following notation. A vector of sizes $(x_1, \ldots, x_n)$ is denoted by z. The vector of first derivatives of the expected adaptation loss with respect to the vector z, evaluated at z^i, is denoted by $W(z^i)$.

$$W(z^i) = \left(\frac{\partial E(L)}{\partial x_1}, \frac{\partial E(L)}{\partial x_2}, \ldots, \frac{\partial E(L)}{\partial x_n} \right)^T \Bigg|_{z = z^i} \tag{4.33}$$

where the symbol T denotes the transpose of a vector.

The matrix of second derivatives, evaluated at z^i, is denoted by $A(z^i)$.

$$A(z^i) = \begin{bmatrix} \dfrac{\partial^2 E(L)}{\partial x_1 \partial x_1} & \dfrac{\partial^2 E(L)}{\partial x_1 \partial x_2} & \cdots & \dfrac{\partial^2 E(L)}{\partial x_1 \partial x_n} \\ \dfrac{\partial^2 E(L)}{\partial x_2 \partial x_1} & & & \vdots \\ \vdots & & & \\ \dfrac{\partial^2 E(L)}{\partial x_n \partial x_1} & \cdots & & \dfrac{\partial^2 E(L)}{\partial x_n \partial x_n} \end{bmatrix}_{z = z^i} \tag{4.34}$$

The matrix A is named the Hessian and is according to (4.12) and (4.13) tri-diagonal. Finally, the vector of sizes for which E(L) reaches a minimum is denoted by z^{opt}, so $W(z^{opt}) = 0$.

Employing Newton's method we first select a vector of starting sizes, z^1. Next $W(z^{opt})$ is expanded in a Taylor series about z^1. Assuming that $W(z^{opt})$ can be closely approximated by a quadratic fit we take only two terms of the expansion into account. This gives

$$W(z^{opt}) = W(z^1) + A(z^1)(z^{opt} - z^1). \tag{4.35}$$

Formula (4.35) can now be solved for z^{opt}, but because of the fact that $W(z)$ in general will not be exactly quadratic the solution of (4.35) is not the optimal one. To obtain a better solution we put the second set of starting values equal to the solution of (4.35) for z^{opt},

$$z^2 = z^1 - A(z^1)^{-1} W(z^1) \tag{4.36}$$

and we repeat the procedure, now with z^2 substituted into (4.35). The iterations are continued in this way until a stop rule terminates the search.

It turned out that the performance of the method is very good if the starting sizes are chosen close to the optimal size pattern. In that case the Hessian will be positive definite and convergence is reached in a few iterations. For a badly chosen starting set, however, convergence can in no way be guaranteed. The major reason for this is that the Hessian, evaluated at such a starting set, might not be positive definite which leads to a step in the wrong direction (i.e. it increases the expected adaptation loss). This happened to be rather the rule than the exception for some problems so that an adjustment of the method became necessary. This adjustment is based on a simplified version of the Marquardt method [Bard, (1974)]. It consists of using,

$$R(z^i) = A(z^i) + \lambda_i B(z^i) \qquad (4.37)$$

where

$$b_{ii} = |a_{ii}| \quad \text{if } a_{ii} \neq 0,$$
$$b_{ii} = 1 \quad \text{if } a_{ii} = 0,$$
$$b_{ij} = 0 \quad \text{if } i \neq j,$$

instead of A in (4.35) and (4.36).

For a sufficiently large value of λ the positive definite matrix λB dominates A so that R is also positive definite. The step $-R^{-1}W$ then will be small but it leads with certainty to a decrease of the expected adaptation loss. If λ is very small, on the other hand, the step $-R^{-1}W$ approximates the Newton step, $-A^{-1}W$.

The following algorithm for the selection of λ_i was used.

1. Select a starting set, z^1. Put i equal to 1 and λ equal to .01. Calculate the E(L) of z^1, say E(L)^1.
2. Calculate $W(z^i)$ and $A(z^i)$.
3. Calculate $R(z^i), R(z^i)^{-1}$, the step, z^{i+1} and E(L)$^{i+1}$.
4. If E(L)$^{i+1}$ < E(L)^i then replace λ with 0.1λ, increase i by 1 and go to step 2.
5. Else, replace λ with 10λ and go to step 3.

As a stop criterion we may use the relative changes of the sizes and/or the relative change of the expected adaptation loss in two successive iterations. Note that both stop criteria ought to be used only after a successful Newton step, thus with λ close to zero.

An often mentioned practical problem when applying Newton's method is the calculation of the Hessian. This generally requires the evaluation of n^2 second derivatives. This objection does not hold for our problem. The Hessian namely is tri-diagonal so that the number of calculations per iteration varies linearly with n instead of quadratically.

As compared to the method of interval bisection Newton's method requires in general more c.p.u. time to find a solution. This is due to the fact that convergence is slow for the large values of λ that often occur if the chosen starting set is not close to the optimal size pattern. We therefore recommend the use of the method of interval bisection if x_{i+1} can be written explicitly as a (simple) function of x_i and x_{i-1}. If this is not the case, or if this function contains a complicated (iterative) evaluation of the inverse cumulative distribution function, Newton's method provides a good alternative.

4.7 THE GRAPHICAL DETERMINATION OF THE OPTIMAL SIZE PATTERN

Besides the two methods discussed in the previous section, it is in a few cases possible to determine the optimal size pattern by means of a graphical method. We will illustrate its use by two examples.

Example 1. In this example it is supposed that the linear loss function is used in combination with shift rule O. Furthermore, we shall assume that the largest size x_n must equal a given constant. This has the obvious effect that the nth recursion formula of (4.22) must be omitted.

According to (4.22), size x_2 has—for given x_1—to satisfy the equation

$$\int_{x_0}^{x_1} h(x)dx = (x_2 - x_1)h(x_1). \qquad (4.38)$$

Selecting a starting value for x_1, say x_1^1, we can determine x_2^1, using the graph of the cumulative distribution function, H, in the following way (see Fig. 4.3). The left-hand side of (4.38) equals the function value of H at x_1^1. This value is equal to the length of the line designated by the letter a in Fig. 4.3. The factor $h(x_1^1)$ on the right-hand side of (4.38) is the slope, α, of the tangent to the curve at x_1^1. If we construct a line, b, with slope α which intersects the x-axis at the point x_1^1 then we may interpret the value of the right-hand side of (4.38) as the length of the perpendicular, c, of a right-angled triangle. Choosing this length equal to $H(x_1)$, it follows that the base of the triangle equals $x_2^1 - x_1^1$. Hence, x_2^1 equals the x-coordinate of the point of intersection of the line $y = H(x_1^1)$ with the line b through x_1^1.

To find $x_3^1, \ldots, x_n^1$ we follow the same procedure as is illustrated in the figure. Next, x_n^1 must be compared with the chosen value of x_n; if x_n^1 equals x_n—within the precision of measurement—the optimal size pattern is found. Otherwise a new starting value, x_1^2, is chosen, depending on the sign of $x_n - x_n^1$ smaller or larger than x_1^1, and the wole process is repeated. Note that for the determination of the successive step sizes $|x_1^j - x_1^{j+1}|$ $(j = 1,2, \ldots)$ the method of interval bisection might be used.

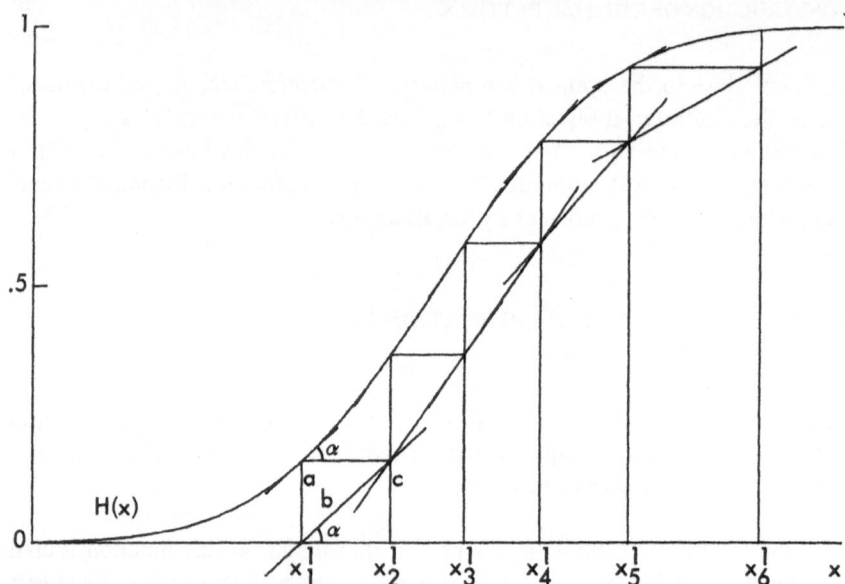

Figure 4.3. The Determination of the Optimal Size Pattern If the Linear Loss Function Is Used in Combination with Shift Rule O

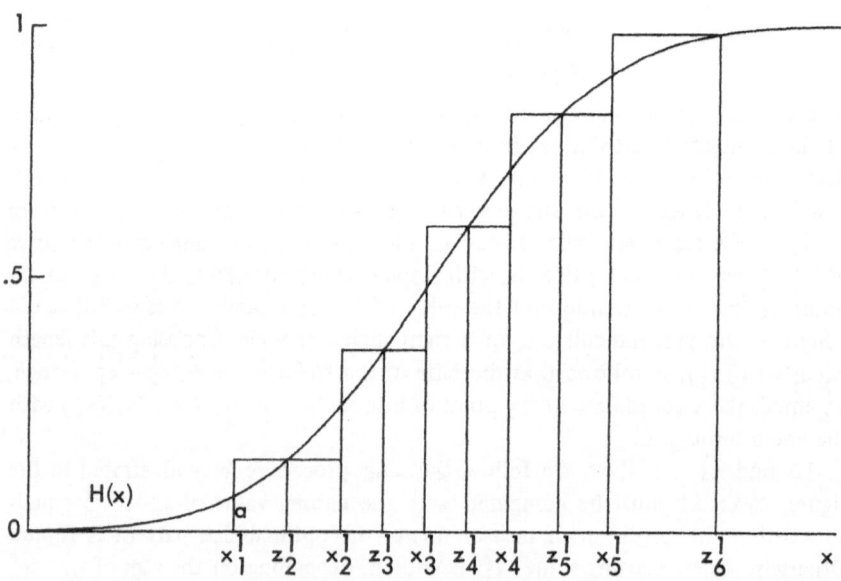

Figure 4.4. The Determination of the Optimal Size Pattern If the Linear Loss Function Is Used in Combination with Shift Rule T

40

Example 2. The optimal size pattern can also be constructed graphically if the linear loss function is used with shift rule T. Putting k_1 equal to k_2 we have according to (4.25),

$$\int_{z_1}^{x_1} h(x)\,dx = \int_{x_1}^{z_2} h(x)\,dx, \qquad (4.39)$$

where by virtue of (4.17), $z_2 = (x_2 + x_1)/2$ and $z_1 = x_0$.

In Fig. 4.4 the cumulative distribution function, $H(x)$, is plotted. Choosing a starting size x_1^1, the left-hand side of (4.39) equals the length of the line indicated in the figure by the letter a. Next ,we extend this line by its own length. Because $H(z_2) = 2H(x_1)$, the y-coordinate of the "top" of the thus obtained line equals $H(z_2^1)$. Consequently, z_2^1 equals the x-coordinate of the intersection point between the line $y = H(z_2^1)$ and the curve $y = H(x)$. Size x_2^1 now easily can be found by measuring the distance between z_2^1 and x_1^1 and adding this to z_2^1. Repeating this procedure we obtain the remaining sizes, $x_3^1, \ldots, x_n^1$. The size pattern $x_1^1, \ldots, x_n^1$ is optimal if the top of the extended line through $(x_n^1, H(x_n^1))$ has a y-coordinate equal to one because this implies that the nth formula of (4.25) is satisfied since z_{n+1}^1 equals z_{n+1}; otherwise a new starting value, x_1^2, is chosen according to the already mentioned rules. In the example k_1 is equal to k_2. It will, however, be clear that only a slight adjustment of the method is necessary to cover also those cases where k_1 does not equal k_2.

4.8   THE DETERMINATION OF THE OPTIMAL VALUE OF n

In the first part of this chapter, the question has been considered how the sizes must be chosen—for a given value of n—such that the expected adaptation loss reaches a minimum. In this section we will discuss some models that can be used to find the value of n which minimizes the (total) average cost per unit of production.

We begin by considering the specific case that the entire expected adaptation loss corresponds with a part of the average production cost. Furthermore, it is assumed that

1. The users of the product apply shift rule O.
2. The production costs per unit can be separated in costs that vary proportionally to the size and costs that are independent of the size. We will call the first type of costs the "incremental" costs and the second type the "initial" costs.

3. The largest size of each size pattern is chosen such that

$$\int_{x_0}^{x_n} h(x)\,dx \text{ approaches 1 } \text{ and } \int_{x_0}^{x_n} x\, h(x)\,dx \text{ approaches } E(x).$$

4. The total sales (in units) are independent of the number of sizes and the size pattern if the latter is optimally chosen.

From the third assumption we see that the loss of not being served is not taken into account. However, as will be made clear in a later part of this section, this loss can easily be incorporated in the models.

According to assumption 2 we may write the production cost of a size x_i as,

$$c(x_i) = ax_i + b \qquad\qquad (4.40)$$

where a denotes the proportionality constant of the incremental costs and b the initial cost. If the cost factors a and b are based on a production of n_1 sizes, the average production cost per unit equal

$$d(n_1) = \sum_{i=1}^{n_1} p_i c(x_i). \qquad\qquad (4.41)$$

The adaptation loss if size x_i is purchased while size x is required reads

$$L(x) = c(x_i) - c(x) = a(x_i - x). \qquad\qquad (4.42)$$

Obviously, the adaptation loss depends only on the cost factor a and not on b. Furthermore, (4.42) indicates that it is not necessary to know the value of a if we want to calculate the optimal size pattern for a given n. Note that the adaptation loss as suffered by the user of the product probably will be larger than the outcome of (4.42). This is caused by the fact that the profit margin and the purchase tax on the product are not taken into account when standardizing a product, since neither form essential cost elements.

By making use of the third assumption, we see further that the minimization of the expected adaptation loss leads to the same size pattern as the minimization of the average production cost because

$$\operatorname*{Min}_{x_1,\ldots,x_{n-1}} \left\{ \sum_{i=1}^{n} p_i\, c(x_i) \right\}$$

$$= \operatorname*{Min}_{x_1,\ldots,x_{n-1}} \left\{ \sum_{i=1}^{n} \int_{x_{i-1}}^{x_i} (a\, x_i + b)\, h(x)\, dx \right\}$$

$$= \underset{x_1, \ldots, x_{n-1}}{\text{Min}} \left\{ \sum_{i=1}^{n} \int_{x_{i-1}}^{x_i} a(x_i - x) h(x) dx + \sum_{i=1}^{n} \int_{x_{i-1}}^{x_i} (ax + b) h(x) dx \right\}$$

$$= a E(x) + \underset{x_1, \ldots, x_{n-1}}{\text{Min}} \{E(L)\} + b. \tag{4.43}$$

The average production cost per unit thus can be split up in three parts. Since the first two terms of (4.43) contain the parameter a, we may interpret these as the average incremental costs of a product. The third term represents the initial costs. We see that the average incremental costs consist of a part that is independent of the number of sizes chosen $(aE(x))$ and a part that depends on n $(E(L))$. Note that if we let n approach infinity and assume that a does not depend on n, the average incremental cost equal $aE(x)$ because the second term of (4.43) vanishes.

After these introductory remarks, we will now study what happens with the average production cost if we let n vary and assume that a and/or b depend on n. As will be clear, the determination of optimal size patterns for varying n will not give rise to any problems since for this calculation none of the parameters a or b have to be known. In this section it will further be assumed that all the size patterns are optimal size patterns; the expected adaptation loss of these size patterns is denoted by $E(L)$ or $E(L)_n$.

To illustrate the calculation of the average production cost per unit, let us suppose that a factory produces n_1 sizes, with the largest size equal to x_{n_1}, and that the production costs per size are given by

$$c_{n_1}(x_i) = a_{n_1} x_i + b_{n_1}. \tag{4.44}$$

The average production costs per unit are then

$$d(n_1) = \sum_{i=1}^{n_1} p_i c_{n_1}(x_i) = a_{n_1} \sum_{i=1}^{n_1} (p_i x_i) + b_{n_1}$$

$$= a_{n_1} E(x) + E(L)_{n_1} + b_{n_1}. \tag{4.45}$$

If now a larger number of sizes n_2 $(n_2 > n_1)$ is chosen, such that $x_{n_2} = x_{n_1}$, then two simultaneous effects occur. In the first place, the expected adaptation loss becomes smaller and according to (4.45) the production cost will decrease. However, in the second place, the production cost will increase because more different sizes are being produced (diseconomies of scale). The latter increase might be caused by an increase of the incremental costs, the initial costs or both.

We will consider first the case that only the initial costs are influenced by a change of n. In that case $a_{n_1} = a_{n_2}$, so that the average production cost per unit can be written as

$$d(n_2) = \sum_{i=1}^{n_2} p_i \, c_{n_2}(x_i) = a_{n_1} \sum_{i=1}^{n_2} p_i x_i + b_{n_2}$$

$$= a_{n_1} E(x) + E(L)_{n_2} + b_{n_2}. \tag{4.46}$$

As a consequence of the increase of n, $E(L)_{n_2}$ in (4.46) will be smaller than $E(L)_{n_1}$ in (4.45). On the other hand, b_{n_2} will be larger than b_{n_1} so that $d(n_2)$ might be smaller than, equal to or larger than $d(n_1)$.

To find the value of n that minimizes the average production cost, we have to determine the relationship between b_n and n. This relation is in general not linear. The increase of the initial cost caused by the adding of an extra size to the size pattern, namely, will in most cases depend on n since for the production of the additional size often some of the existing facilities can be used. In general, good approximations can be obtained by making use of the function [Sittig, (1972)],

$$b_n = en^\epsilon \tag{4.47}$$

to describe the relation between b_n and n. A function of this type is for instance common in inventory problems. According to the formula of Wilson [Wagner, (1969)], a doubling of the number of sizes, which has the same effect as halving the average series length, increases the inventory costs per unit with a factor $\sqrt{2}$. Hence, the value of ϵ equals $1/2$ in this case.

Some numerical experimentation with the model discussed in Wagner showed that the value of ϵ will be slightly less than $1/2$ in case the demand is assumed to be probabilistic rather than deterministic and the lead time differs from zero. The introduction of safety stocks, which in general is necessary in the latter situation, has therefore not much influence on the value of ϵ.

After the estimation of e and ϵ it follows from (4.46) that the optimal n can be found by minimizing the expression:

$$E(L)_n + en^\epsilon \tag{4.48}$$

where $E(L)_n$ denotes the expected adaptation loss of an optimal size pattern consisting of n sizes.

Having considered the case that only the initial costs were influenced by n, we will now discuss the more realistic case that both a and b are increasing functions of n. If we suppose that—when producing n_1 sizes—the production cost of size x_i and the average production cost per unit are given by respectively (4.44) and (4.45) then at a production of n_2 ($n_2 > n_1$) sizes these relations become

$$c_{n_2}(x_i) = a_{n_2} x_i + b_{n_2} \tag{4.49}$$

and

$$d(n_2) = \sum_{i=1}^{n_2} p_i c_{n_2}(x_i) = a_{n_2} \sum_{i=1}^{n_2} (p_i x_i) + b_{n_2}$$

$$= a_{n_2} E(x) + E(L)_{n_2} + b_{n_2}. \qquad (4.50)$$

As compared with (4.45) the first and third term of (4.50) have become larger due to the larger values of a_{n_2} and b_{n_2}. The second term of (4.50) equals

$$E(L)_{n_2} = (a_{n_2}) \left(\sum_{i=1}^{n_2} \int_{x_{i-1}}^{x_i} (x_i - x) h(x) dx \right). \qquad (4.51)$$

Clearly, the first factor in (4.51) is larger than the corresponding factor a_{n_1} in $E(L)_{n_1}$ while the second factor is smaller. The overall effect of an increase of n on the expected adaptation loss thus cannot be determined without knowing the actual values of the variables. Note that this is opposed to the behavior of the expected adaptation loss in the first model; an increase of n always led there to a decrease of $E(L)$.

In the specific case that both a_{n_1} and b_{n_1} increase with the same percentage, say z, if n_2 instead of n_1 sizes are chosen, (4.50) may be written as

$$d(n_2) = \sum_{i=1}^{n_2} p_i c_{n_2}(x_i) = \left(\frac{100 + z}{100} \right) \left(a_{n_1} \sum_{i=1}^{n_2} p_i x_i + b_{n_1} \right). \qquad (4.52)$$

From (4.52) we see that in the second factor the parameters a_{n_1} and b_{n_1}, which were based on a production of n_1 sizes, are used. By means of the first factor, the result of the second factor is corrected for the change in a_{n_1} and b_{n_1}. To determine the optimal value of n, it is necessary to find the functional relationship between the first factor and n. A good fit was obtained in a number of practical problems by the use of a simple function similar to (4.47). This function has the form $(n/n_1)^\epsilon$ so that the average production cost reads

$$d(n) = \left(\frac{n}{n_1} \right)^\epsilon \left(a_{n_1} \sum_{i=1}^{n} p_i x_i + b_{n_1} \right). \qquad (4.53)$$

The parameter ϵ in (4.53) varies between zero and one and for the estimation of ϵ all types of costs that are influenced by n (for instance inventory costs, set-up costs) are taken into consideration. According to Sittig (1969) the parameter ϵ might be interpreted as the degree of mechanization in a given branch of industry. If the value of ϵ is close to zero, the industry possesses a low degree of mechanization; traditional building might furnish an example. On the other hand, industries with a high degree of mechanization are characterized by a larger value of ϵ. For most industries the value of ϵ varies between .05 and .35. In a casting room of a ready-made clothing shop, for example, ϵ was found to be equal to .08 while in a transistor factory it was equal to .33. Some other empirical values of ϵ that were assembled by Sittig are presented in Table 4.1 [Sittig, (1978)]. If the value of ϵ has been estimated, the value of n that minimizes expression (4.53) can easily be found by calculating d(n) for a range of values of n. An application of this model can be found in a report on the standardization of the thickness of window panes [AKB, (1976)].

In the models discussed above, it was assumed that the adaptation losses correspond entirely with a part of the production cost and that the production costs per unit are a linear function of the size. Furthermore, the loss of not being served was not taken into consideration. However, the applicability of the models can easily be increased by changing the assumptions 2 and 3, that were made in the beginning of this section, into

2. The production cost per unit can be separated into incremental and initial costs. The incremental costs are supposed to be a cost factor times a function of the size.
3. The part of the adaptation loss that is not included in the production cost is proportional to the part that is included in the production cost.

From assumption 2 it follows that the production cost of a size x_i can be written as

$$c(x_i) = at(x_i) + b \qquad (4.54)$$

Table 4.1. Empirical Values of ϵ

Product	Country	ϵ
Floor joists	Spain	.077
Dwellings	Netherlands	.084
Prefabricated concrete panels	Ukrainian, SSR	.088
Timber windows	Germany	.101
Ethylene	U.K.	.130
Small tools	U.K.	.141
Standard tools	U.K.	.213
Integrated circuits	U.K.	.350

where b again denotes the initial costs. According to (4.54) and assumption 3, the adaptation loss equals

$$L(x) = c(x_i) - c(x) + \alpha(c(x_i) - c(x)) = a(1 + \alpha)(t(x_i) - t(x)) \qquad (4.55)$$

where $\alpha(c(x_i) - c(x))$ denotes the part of the adaptation loss that does not correspond with any part of the production cost.

If x is larger than x_n, the consumer cannot be served. Since the loss of not being served can be interpreted as the extra cost, B, he has to pay for a specially constructed product, we may put the production cost of such a product equal to

$$c(x) = at(x) + b + B \quad \text{if } x > x_n. \qquad (4.56)$$

The total average cost per product then reads

$$d(n) = \sum_{i=1}^{n} p_i c(x_i) + \sum_{i=1}^{n} \alpha \int_{x_{i-1}}^{x_i} (c(x_i) - c(x))h(x)dx$$

$$+ \int_{x_n}^{x_{n+1}} (a\,t(x) + b + B)h(x)dx \qquad (4.57)$$

The minimization of $d(n)$ again leads to the same size pattern as the minimization of the expected adaptation loss because

$$\operatorname*{Min}_{x_1, \ldots, x_n} \{d(n)\} = a\,E(t(x)) + b$$

$$+ \operatorname*{Min}_{x_1, \ldots, x_n} \left\{ a(1 + \alpha) \sum_{i=1}^{n} \int_{x_{i-1}}^{x_i} (t(x_i) - t(x))h(x)dx \right.$$

$$\left. + B \int_{x_n}^{x_{n+1}} h(x)dx \right\}$$

$$= a\,E(t(x)) + b + \operatorname*{Min}_{x_1, \ldots, x_n} \{E(L)\}. \qquad (4.58)$$

It will be clear that the first model, given by (4.48), can be used without modifications to find the optimal value of n. When using the second model, we have to realize that besides b, the factor a also depends on n. This implies that— according to (4.58)—the relationship between a and n must be known before the optimal size patterns can be calculated for varying n. The determination of the optimal n further goes exactly as described above.

5 THE RECURSION FORMULAE FOR THE HOMOGENEOUS DISTRIBUTION

5.1 INTRODUCTION

In the previous chapter the recursion formulae were derived for the loss functions given in Table 2.1. The probability density function of the required sizes was not yet specified in these formulae. In the Chaps. 5, 6, 7 and 8 we shall substitute successively some well-known density functions into the recursion formulae. This has the effect that in most cases the recursion formulae and the formula of the expected adaptation loss can be reduced to a more simple form.

The following four probability density functions will be taken into consideration:

1. the homogeneous distribution,
2. the normal distribution,
3. the log normal distribution,
4. the gamma distribution.

We have included the homogeneous distribution because of its simple form. As we will show, when using this distribution the recursion formulae can be solved analytically if the linear loss function is used and/or shift rule T is applied. Except for the exponential distribution—which is a special case of the gamma distribution—the homogeneous distribution is the only distribution which permits an analytical solution. The other three distributions merit their inclusion from the fact that they frequently are used in practice to approximate empirical distributions of the demand for product properties.

5.2 THE PROBABILITY DENSITY FUNCTION

The homogeneous distribution is defined on the interval (a,b) with $a \geqslant 0$. The probability density function is

$$h(x) = 1/(b - a) \quad 0 \leqslant a \leqslant x \leqslant b,$$
$$= 0 \quad \text{elsewhere.} \tag{5.1}$$

Note that when using this distribution x_0 equals a and x_{n+1} equals b. The mean and variance of the distribution are

$$E(x) = (a + b)/2, \quad V(x) = (b - a)^2/12. \tag{5.2}$$

5.3 THE NONLINEAR LOSS FUNCTION IN COMBINATION WITH SHIFT RULE O

By virtue of (4.29) we have

$$x_{i+1} = x_i \{1 + \alpha (1 - x_{i-1}/x_i)\}^{1/\alpha} \quad i = 1, \dots, n - 1. \tag{5.3}$$

Furthermore, according to the nth formula of (4.28), sizes x_{n-1} and x_n have to satisfy the relation

$$x_n^{\alpha-1}(x_n - x_{n-1}) = B/k\alpha. \tag{5.4}$$

However, when using the homogeneous distribution, it may occur that no sizes x_n and x_{n-1} satisfying (5.4) can be found. For instance, if n = 1, the left-hand side of (5.4) equals maximally $b^{\alpha-1} (b - a)$ and therefore no solution of (5.4) exists if the right-hand side is larger than this value. This problem is caused by the fact that the recursion formulae were derived without taking into account the restrictions

$$x_0 \leqslant x_1 \leqslant x_2 \dots \leqslant x_n \leqslant x_{n+1}. \tag{5.5}$$

These restrictions, however, can be easily incorporated in the model by using Lagrange multipliers. This leads to the result that x_n must be chosen equal to $b (= x_{n+1})$ if no solution of (5.4) exists.

The expected adaptation loss equals:

$$E(L) = k \sum_{i=1}^{n} \left\{ \int_{x_{i-1}}^{x_i} \frac{x_i^{\alpha} - x^{\alpha}}{b - a} dx \right\} + B \int_{x_n}^{b} \frac{1}{b - a} dx$$

$$= \frac{k}{b - a} \left[\sum_{i=1}^{n} \{x_i^{\alpha}(x_i - x_{i-1})\} + \frac{a^{\alpha+1} - x_n^{\alpha+1}}{\alpha + 1} \right] + B \frac{b - x_n}{b - a}. \tag{5.6}$$

The method of interval bisection is used to find the optimal size pattern. The position of the optimal size pattern is heavily influenced by the value of α. This is illustrated by Table 5.1 which contains the optimal size patterns for several values of α. The other parameter values were respectively

$$a = 0, \quad k = 1, \quad n = 5,$$
$$b = 10, \quad B = 10.$$

When comparing the size patterns we see that the sizes—except x_5—become larger if α increases from 1 to 1.5. Furthermore, the distance between two successive sizes, x_{i+1} and x_i, is equal to two for all i if $\alpha = 1$. However, for larger values of α this distance is a decreasing function of i. Loosely speaking, this is due to the fact that—if we let α increase and keep the size pattern fixed—the expected adaptation loss of the ith subpopulation grows faster than that of the $(i - 1)$th subpopulation $(i = 2, \ldots, n)$. Consequently, the expected adaptation loss of the total population can be reduced by choosing the service areas of the larger sizes smaller than those of the smaller sizes.

The table also shows that there exists a value of α between 1.5 and 1.75, say α^*, for which the nth recursion formula (5.4) is satisfied. The expected loss that arises if the consumer is served becomes—as compared to the expected loss due

Table 5.1. Some Optimal Size Patterns for Varying α (Nonlinear Loss Function, Shift Rule O)

α	E(L)	x_1	x_2	x_3	x_4	x_5
1.00	1.00	2.00	4.00	6.00	8.00	10.00
1.25	1.82	2.25	4.31	6.27	8.16	10.00
1.50	3.24	2.49	4.59	6.50	8.29	10.00
1.75	5.08	2.10	3.74	5.19	6.50	7.74
2.00	6.14	1.69	2.93	3.98	4.92	5.79
5.00	8.44	0.88	1.26	1.51	1.71	1.87

to not being served—for increasing values of α beyond α^* that large that it will be more economical to reduce the coverage than to let the sizes become larger. Therefore the whole size pattern shifts to the left for α larger than α^*.

Finally, if α approaches infinity all the sizes approach unity. From (5.6) it then follows that $E(L)$ approaches 9.

5.4.1 THE LINEAR LOSS FUNCTION IN COMBINATION WITH SHIFT RULE O

Substituting $\alpha = 1$ into (5.3) and (5.4) we obtain

$$x_i - x_{i-1} = x_{i+1} - x_i \quad i = 1, \ldots, n - 1, \tag{5.7}$$

$$x_n - x_{n-1} = B/k. \tag{5.8}$$

According to (5.7) the difference between two successive sizes equals a constant, say r. Size x_i therefore equals

$$x_i = a + ir \quad i = 1, \ldots, n. \tag{5.9}$$

Since x_n ($= a + nr$) cannot exceed b, the constant r must be smaller than or equal to $(b - a)/n$. This leads to

$$r = (b - a)/n \quad \text{if } B/k \geqslant (b - a)/n, \tag{5.10}$$

$$r = B/k \quad \text{if } B/k < (b - a)/n. \tag{5.11}$$

The expected adaptation loss is found by putting $\alpha = 1$ and substituting (5.7) and (5.8) into (5.6)

$$E(L) = (\tfrac{1}{2}knr^2 + B(b - a - nr))/(b - a). \tag{5.12}$$

Using (5.10) and (5.11) we may reduce this expression to

$$E(L) = \tfrac{1}{2}k(b - a)/n \quad \text{if } B/k \geqslant (b - a)/n, \tag{5.13}$$

$$E(L) = B - nB^2/(2k(b - a)) \quad \text{if } B/k < (b - a)/n. \tag{5.14}$$

In the rest of this section we assume the value of B/k to be larger than $(b - a)$. The value of r is then given by (5.10) for all values of n. Hence, the probabilities of the service areas are equal to $p_i = 1/n$ for all i.

If the sizes are chosen optimally it follows that the expected adaptation loss of the ith subpopulation equals

$$k \int_{x_{i-1}}^{x_i} (x_i - x)/(b - a)\, dx = \tfrac{1}{2}k(b - a)/n^2. \tag{5.15}$$

The index i does not appear in (5.15) which means that every subpopulation must have the same expected adaptation loss. The conditional expected adaptation loss of the ith subpopulation is

$$k \int_{x_{i-1}}^{x_i} (x_i - x)/(b - a)dx/p_i = \frac{1}{2}k(b - a)/n. \qquad (5.16)$$

Clearly, also the conditional expected adaptation loss is the same for every subpopulation.

We thus have the rather remarkable fact that—when using the homogeneous distribution and the linear loss function with shift rule O—the optimal size pattern satisfies all the criteria discussed in Chap. 3. These were:

1. minimization of the expected adaptation loss,
2. equal total expected adaptation loss per service area,
3. equal conditional expected adaptation loss per service area,
4. equal sales per size.

Furthermore, the size pattern forms an arithmetical series; it therefore also satisfies one of the rules which are used in practice to select the sizes.

5.4.2 THE CHOICE OF THE OPTIMAL VALUE OF n

Having considered the choice of the optimal size pattern for a given n, we can compute next the optimal value of n by means of one of the models discussed in Sec. 4.8.

Let us suppose first that the initial costs, say g, depend on n whereas the proportionality constant of the incremental costs, k, is not influenced by n. The average production costs per unit then can be written as (see 4.46),

$$d(n) = kE(x) + E(L)_n + g_n \qquad (5.17)$$

where $E(L)_n$ is given by (5.13). From (5.13) we see that the expected adaptation loss decreases with increasing n. However, the percentage reduction of the expected adaptation loss for two successive values of n becomes smaller for larger n. At a certain value of n, the reduction of the expected adaptation loss will be smaller than the rise of the initial costs per unit if n is increased by one. Clearly, this value of n is the optimal value.

Since the initial costs per unit are assumed to be of the form (see 4.47),

$$g_n = en^\epsilon \qquad (5.18)$$

the optimal value of n can be found by differentiating the expression,

$$d(n) = k(b + a)/2 + \frac{1}{2}k(b - a)/n + en^\epsilon \qquad (5.19)$$

with respect to n,

$$\partial d(n)/\partial n = -\tfrac{1}{2}k(b - a)/n^2 + e\epsilon n^{\epsilon-1}. \tag{5.20}$$

Putting (5.20) equal to zero we have

$$n = \{\tfrac{1}{2}k(b - a)/e\epsilon\}^{1/(\epsilon+1)} \tag{5.21}$$

The factor $\tfrac{1}{2}k(b - a)$ in (5.21) represents the expected adaptation loss if $n = 1$ and the factor e equals the initial costs per unit, also if $n = 1$. Denoting the ratio of these factors by q and taking into consideration the fact that the optimal value of n cannot be smaller than unity, we obtain

$$n^{opt} = (q/\epsilon)^{1/(\epsilon+1)} \quad \text{if } 0 \leqslant \epsilon < q,$$
$$n^{opt} = 1 \qquad\qquad \text{if } 1 \geqslant \epsilon \geqslant q. \tag{5.22}$$

The average production costs per unit for $n = n^{opt}$ are equal to

$$d(n^{opt}) = kE(x) + e(1 + \epsilon)(q/\epsilon)^{\epsilon/(\epsilon+1)} \quad \text{if } 0 \leqslant \epsilon < q,$$
$$d(n^{opt}) = d(1) = kE(x) + e(1 + q) \qquad \text{if } 1 \geqslant \epsilon \geqslant q. \tag{5.23}$$

Figure 5.1 shows the integer values of n^{opt} as a function of q and ϵ. As can be seen, n^{opt} is, for fixed q, a decreasing function of the degree of mechanization, ϵ. Furthermore, it follows from (5.22) that n^{opt} approaches infinity if ϵ approaches zero. The reason is that for small ϵ the initial costs per unit hardly are influenced by the number of different sizes produced which implies that the expected adaptation loss can be made arbitrarily small by taking n large. The average production costs per unit than approach $kE(x) + e$.

Another assumption we can make – when deriving the optimal value of n – is that the initial costs, e, and the proportionality constant of the incremental costs, k, both depend on n. In the specific case that e and k increase with the same percentage if n is raised by a certain amount we have (see 4.53),

$$d(n) = (n/n_1)^{\epsilon} (kE(x) + E(L)_n + e) \tag{5.24}$$

where k and e are based on production of n_1 different sizes. The derivative of $d(n)$ with respect to n equals

$$\partial d(n)/\partial n = (n^{\epsilon-1}/n_1^{\epsilon}) \{\epsilon kE(x) + e\epsilon + (\epsilon - 1)\tfrac{1}{2}k(b - a)/n\}. \tag{5.25}$$

Putting (5.25) equal to zero and solving for n leads to

$$n^{opt} = \frac{(1 - \epsilon)(\tfrac{1}{2}k(b - a))}{\epsilon(kE(x) + e)} \quad \text{if } \epsilon < \frac{\tfrac{1}{2}k(b - a)}{\tfrac{1}{2}k(b - a) + kE(x) + e},$$

$$n^{opt} = 1 \qquad\qquad \text{if } \epsilon \geqslant \frac{\tfrac{1}{2}k(b - a)}{\tfrac{1}{2}k(b - a) + kE(x) + e}. \tag{5.26}$$

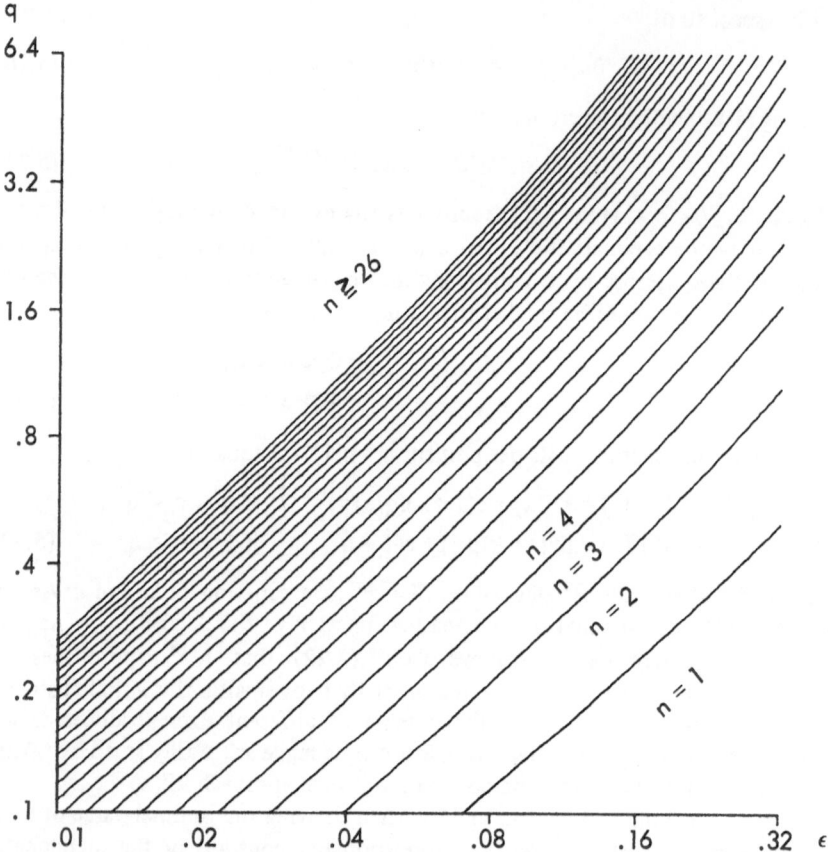

Figure 5.1. The Optimal Value of n as a Function of q and ε

As in the first model, n^{opt} is a decreasing function of ϵ. Also, if ϵ approaches zero, the optimal value of n approaches infinity. However, as opposed to the result of the first model, n^{opt} now also depends on the mean of x. Finally, if n is chosen optimally, the average production costs per unit can be written as

$$d(n^{opt}) = \left(\frac{\frac{1}{2}k(b-a)}{\epsilon n_1}\right)^{\epsilon}\left(\frac{kE(x)+e}{1-\epsilon}\right)^{1-\epsilon} \quad \text{if } \epsilon < \frac{\frac{1}{2}k(b-a)}{\frac{1}{2}k(b-a)+kE(x)+e},$$

$$d(n^{opt}) = d(1) = \frac{1}{n_1^{\epsilon}}(kE(x)+\frac{1}{2}k(b-a)+e) \quad \text{if } \epsilon \geq \frac{\frac{1}{2}k(b-a)}{\frac{1}{2}k(b-a)+kE(x)+e}. \quad (5.27)$$

5.5 THE NONLINEAR LOSS FUNCTION IN COMBINATION WITH SHIFT RULE T

After substitution of $\alpha = 1$ and $h(x) = 1/(b - a)$ we can reduce (4.15) to

$$x_i - x_{i-1} = x_{i+1} - x_i \quad i = 2, \ldots, n - 1, \tag{5.28}$$

$$x_1 - a = \gamma(x_2 - x_1), \tag{5.29}$$

$$(1 - \gamma)(x_n - x_{n-1}) = b - x_n, \tag{5.30}$$

where

$$\gamma = \sqrt[\beta]{k_1}/(\sqrt[\beta]{k_1} + \sqrt[\beta]{k_2}). \tag{5.31}$$

The recursion formulae can be solved in the following way. First note that according to (5.28) the distance between two sizes x_{i-1} and x_i ($i = 2, \ldots, n$) equals a constant which we denote by r. Secondly, the sum of the distances,

$$\sum_{i=1}^{n+1} (x_i - x_{i-1}) \tag{5.32}$$

obviously equals $b - a$. Using (5.29) and (5.30), we thus may write (5.32) as,

$$\sum_{i=1}^{n+1} (x_i - x_{i-1}) = b - a = \gamma r + (n - 1)r + (1 - \gamma)r = nr \tag{5.33}$$

which gives

$$r = (b - a)/n. \tag{5.34}$$

Combining (5.34) with (5.28), (5.29) and (5.30) we obtain

$$x_i = a + ((b - a)/n)(i + \gamma - 1) \quad i = 1, \ldots, n. \tag{5.35}$$

Furthermore, by substituting the result (5.35) into (4.17)

$$z_i = a + ((b - a)/n)(i - 1) \quad i = 2, \ldots, n. \tag{5.36}$$

Finally, as readily can be verified, the expected adaptation loss of a size pattern chosen according to (5.35) equals

$$E(L) = \frac{k_1 k_2}{(\beta + 1)} \left[\frac{b - a}{n(\sqrt[\beta]{k_1} + \sqrt[\beta]{k_2})} \right]^{\beta}. \tag{5.37}$$

The results if the loss function is linear can easily be derived from the foregoing formulae by putting β equal to unity. An example is given by Fig. 5.2 which shows the optimal size pattern obtained by using the linear loss function in combination with the following parameter values

$$k_1 = 3, \quad a = 0, \quad n = 4,$$
$$k_2 = 1, \quad b = 1.$$

The expected adaptation loss of this size pattern equals — as can be read from the figure — $3/32$.

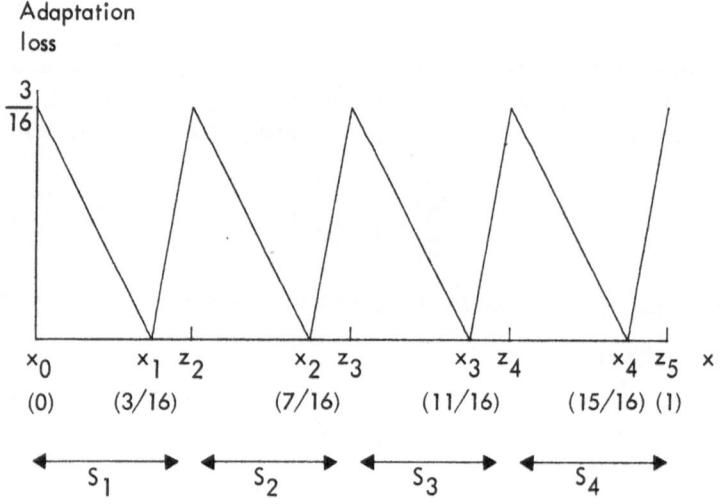

Figure 5.2. The Optimal Size Pattern and the Adaptation Loss as a Function of the Required Size

6 THE RECURSION FORMULAE FOR THE NORMAL DISTRIBUTION

6.1 THE PROBABILITY DENSITY FUNCTION

The probability density function of the normal distribution has the form

$$h(x) = \frac{1}{\sigma\sqrt{2\pi}} e^{-((x-\mu)/\sigma)^2/2} \qquad -\infty < x < \infty. \tag{6.1}$$

The mean and standard deviation are respectively μ and σ. Furthermore, according to (4.1) and (4.2), x_0 equals 0 and x_{n+1} equals ∞. Referring to the remark concerning the normal distribution in Sec. 4.2, it is assumed that the ratio σ/μ is that small that the integrals

$$\int_{x_0}^{x_{n+1}} h(x)\,dx, \qquad \int_{x_0}^{x_{n+1}} x\,h(x)\,dx \quad \text{and} \quad \int_{x_0}^{x_{n+1}} (x-\mu)^2\,h(x)\,dx,$$

are, without noticeable error, equal to respectively unity, μ, and σ^2. In the sequel we often will use the transformation,

$$v = (x - \mu)/\sigma \tag{6.2}$$

and the substitutions

$$v_i = (x_i - \mu)/\sigma \quad i = 0, \ldots, n+1. \tag{6.3}$$

The variable v has the standard normal density for which the letter f is reserved. The letter F then denotes the cumulative standard normal distribution.

The variables $v_1, \ldots, v_n$ are introduced to simplify the recursion formulae in some cases. We call these variables the transformed sizes; the set $(v_1, \ldots, v_n)$ is called the transformed size pattern.

6.2 THE NONLINEAR LOSS FUNCTION IN COMBINATION WITH SHIFT RULE O

The optimal size pattern might—according to (4.29)—be generated with the method of interval bisection. To calculate p_i (i = 1, ..., n) the numerical approximation of the normal integral as given by Abramowitz (1972) can be used. We further mention that, when using the normal distribution, there always exists a solution of the nth equation of (4.28), because the largest size x_n can be made arbitrarily large.

6.3 THE LINEAR LOSS FUNCTION IN COMBINATION WITH SHIFT RULE O

We first will pay some attention to the specific case that n = 1. According to (4.22) size x_1 must satisfy the condition

$$\int_{x_0}^{x_1} h(x)dx = (B/k)h(x_1). \tag{6.4}$$

Applying (6.2) and (6.3) to (6.4) we find

$$F(v_1) = (B/k\sigma)f(v_1). \tag{6.5}$$

The expected adaptation loss equals

$$\begin{aligned}
E(L) &= k \int_{x_0}^{x_1} (x_1 - x) h(x)dx + B \int_{x_1}^{\infty} h(x)dx \\
&= k\sigma \int_{v_0}^{v_1} (v_1 - v) f(v)dv + B \int_{v_1}^{\infty} f(v)dv.
\end{aligned} \tag{6.6}$$

Using (B.12) and (B.13) in Appendix B, we may reduce (6.6) to

$$E(L) = k\sigma(v_1 F(v_1) + f(v_1)) + B(1 - F(v_1)). \tag{6.7}$$

In Fig. 6.1, $E(L)$ is plotted as a function of v_1 for the parameter values $k = \sigma = 1$ and $B = 10$. From (6.7) we see that $E(L)$ approaches $k\sigma v_1$ if v_1 is large. If, on the other hand, v_1 assumes a large negative value, then $E(L)$ is approximately equal to B. The first derivative of $E(L)$ with respect to v_1 is plotted in Fig. 6.2.

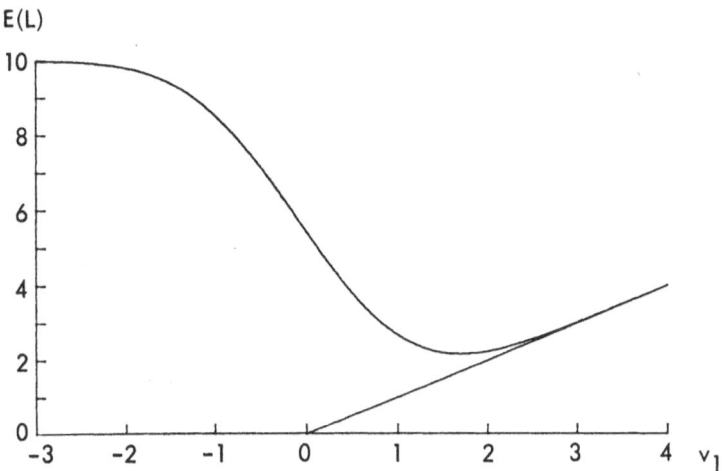

Figure 6.1. The Expected Adaptation Loss as a Function of v_1

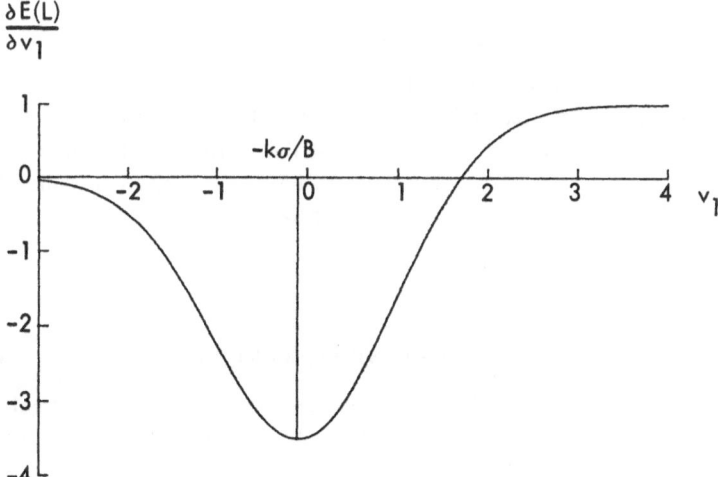

Figure 6.2. The First Derivative of $E(L)$ with Respect to v_1

This function shows why the Marquardt adjustment to Newton's method is necessary. The function has its minimum at $v_1 = -k\sigma/B$. If the starting size v_1^1 for example is chosen smaller than $-k\sigma/B$, the tangency line at v_1^1 will cross the x-axis at a point v_1^2 which is smaller than v_1^1. Continuing the iterations, v_1^j ($j = 3, \ldots$) becomes smaller and smaller; for the chosen starting value the procedure thus clearly does not converge to the optimal value of v_1, say v_1^{opt}. In the simple case that $n = 1$ a rather sharp upper and lower bound can be derived for v_1^{opt}. We proceed as follows,

$$(1/k\sigma)(\partial E(L)/\partial v_1) = F(v_1) - (B/k\sigma)f(v_1) \tag{6.9}$$

which implies

$$(1/k\sigma)(\partial E(L)/\partial v_1) < 1 - (B/k\sigma)f(v_1). \tag{6.10}$$

The right-hand side of (6.10) equals zero for a smaller value of v_1 than the left-hand side, hence

$$v_1^{opt} > \sqrt{-2\ln(\sqrt{2\pi}\, k\sigma/B)} \quad \text{if } B/k\sigma > \sqrt{2\pi}\, (= 2.51). \tag{6.11}$$

We also have

$$1 - F(v_1) < f(v_1) \quad \text{if } v_1 > .303. \tag{6.12}$$

According to (6.11), v_1^{opt} is larger than .303 if $B/k\sigma$ is larger than 2.625. Assuming the latter being the case we obtain from (6.9) and (6.12),

$$(1/k\sigma)(\partial E(L)/\partial v_1) > 1 - f(v_1) - (B/k\sigma)f(v_1) \tag{6.13}$$

which implies

$$v_1^{opt} < \sqrt{-2\ln(\sqrt{2\pi}\, k\sigma/(k\sigma + B))} \quad \text{if } B/k\sigma > 2.625. \tag{6.14}$$

For the parameter values $k = \sigma = 1$ and $B = 10$, v_1^{opt} is by virtue of (6.11) and (6.14) a point in the interval (1.66, 1.72). The relative difference between these bounds is 3.4% which decreases very fast if the factor $B/k\sigma$ becomes larger. If $B/k\sigma$ for instance equals 100 the relative difference is .13%. The value of $E(L)$ for large $B/k\sigma$ and v_1 optimally chosen is approximately,

$$E(L) \simeq k\sigma(v_1^{opt} + 1/v_1^{opt}) \tag{6.15}$$

which follows from the fact that

$$F(v) \simeq 1 - (1/v)f(v) \quad \text{for large } v. \tag{6.16}$$

Having considered the case that only one size was to be chosen, we next take n larger than one. Applying (6.2) and (6.3) to (4.22) we obtain the recursion formulae,

$$F(v_i) - F(v_{i-1}) = (v_{i+1} - v_i) f(v_i) \qquad i = 1, \dots, n - 1,$$

$$F(v_n) - F(v_{n-1}) = (B/k\sigma)f(v_n) \tag{6.17}$$

which can be solved with the method of interval bisection. The expected adaptation loss equals,

$$E(L) = k \sum_{i=1}^{n} \int_{x_{i-1}}^{x_i} (x_i - x) \, h(x) dx + B \int_{x_n}^{\infty} h(x) dx \tag{6.18}$$

which can be reduced to

$$E(L) = k\sigma \left[\sum_{i=1}^{n} \left(v_i \int_{v_{i-1}}^{v_i} f(v) dv \right) + f(v_n) \right] + B(1 - F(v_n)). \tag{6.19}$$

To illustrate the position of the transformed sizes, the optimal transformed size patterns are calculated for $B/k\sigma = 10$ and n ranging from 1 to 5. These size patterns are plotted in Fig. 6.3.

Several points are to be mentioned. First, the size patterns (with $n > 2$) certainly do not form an arithmetical series as was the case when using the homogeneous distribution with the linear loss function. Taking for instance the size pattern with $n = 5$ we see that the differences between the successive sizes — instead of being equal to each other — heavily depend on the height of the density function; near the mode of the distribution these differences are smaller than at the tails. The position of the optimal size pattern is thus closely related to the shape of the density function. Standardizing a product property without taking into account the probability density function — as unfortunately often is done in practice — therefore generally will lead to unnecessary losses. In Chap. 12 we show with some examples how large these losses may become.

Another interesting point is that — contrary to what one might expect — the sizes are not distributed symmetrically around the mode. As namely can be seen from the figure, most sizes of each size pattern are located to the right of the mode. The reason for this is the following. If we take a service area which is located to the left of the mode, for instance S_1 in the figure with $n = 5$, it will be clear that the adaptation loss of a consumer whose demand is in the left part of this service area is larger than that of a consumer with a demand in the right part. The density function, however, is increasing above S_1 which means that — when calculating the expected adaptation loss of S_1 — the larger adaptation losses in the left part of S_1 are weighted by smaller "probabilities" than the smaller losses in the right part of S_1. For service areas to the right of the mode the reverse is true because the density function decreases above these service areas.

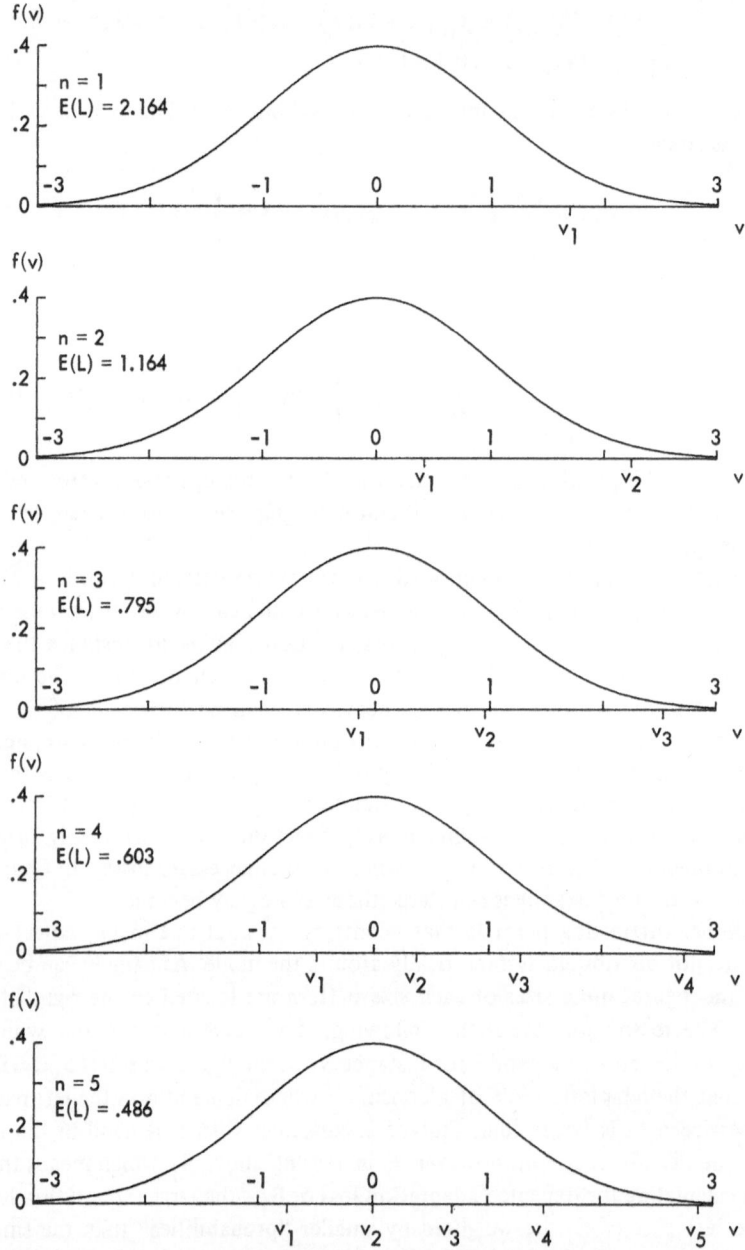

Figure 6.3. Optimal Transformed Size Pattern with $B/k\sigma = 10$ (Normal Distribution, Linear Loss Function and Shift Rule O)

Choosing a size pattern symmetrically around the mode therefore implies that the expected adaptation loss of the subpopulation of consumers with a demand below the mode will be smaller than that of the subpopulation with a demand above the mode. This leads to the idea that the (total) expected adaptation loss of the symmetrical size pattern probably can be reduced by selecting more sizes above the mode. For the given example the optimal solution shows that this was indeed the case. Finally, the optimal size patterns are calculated for a number of different values of $B/k\sigma$ and n. These size patterns are listed in Table A.1. of Appendix A. The table also contains the expected adaptation loss divided by B as well as the coverage of each size pattern.

6.4 THE QUADRATIC LOSS FUNCTION IN COMBINATION WITH SHIFT RULE T

We have, according to (4.15), the recursion formulae

$$k_2 \int_{z_i}^{x_i} (x_i - x)^{\beta-1} h(x)\, dx$$

$$= k_1 \int_{x_i}^{z_{i+1}} (x - x_i)^{\beta-1} h(x)\, dx \qquad i = 1, \ldots, n. \qquad (6.20)$$

To calculate the value of (6.20) for an arbitrary value of β we have to use the technique of numerical integration except in those few cases where the integrals can be evaluated analytically. In this study we restrict ourselves to some of the latter cases. Putting β equal to two we have after transformation (6.2) and substitutions (6.3)

$$k_2 \int_{z_i^*}^{v_i} (v_i - v)\, f(v)\, dv$$

$$= k_1 \int_{v_i}^{z_{i+1}^*} (v - v_i)\, f(v)\, dv \qquad i = 1, \ldots, n \qquad (6.21)$$

where

$$z_i^* = (z_i - \mu)/\sigma \qquad i = 1, \ldots, n+1. \qquad (6.22)$$

The method of interval bisection is not very well suited to solve the set of equations (6.21) as v_{i+1} cannot be written explicitly as a function of v_i and v_{i-1}. Therefore we apply Newton's method. The Hessian, which can be derived from (6.21), has the form

$$a_{ii} = k_2 \int_{z_i^*}^{v_i} f(v)\,dv + k_1 \int_{v_i}^{z_{i+1}^*} f(v)\,dv$$

$$- \frac{\sqrt{k_1 k_2}}{(\sqrt{k_1} + \sqrt{k_2})^2} \{k_2(v_i - v_{i-1})\, f(z_i^*)$$

$$+ k_1(v_{i+1} - v_i)\, f(z_{i+1}^*)\} \quad i = 1, \ldots n,$$

$$a_{i,i+1} = a_{i+1,i} = \frac{-k_1 k_2}{(\sqrt{k_1} + \sqrt{k_2})^2}(v_{i+1} - v_i)\, f(z_{i+1}^*) \quad i = 1, \ldots, n-1 \tag{6.23}$$

and all other elements equal zero. As can be seen, the function and integral evaluations in (6.23) are exactly the same as those that were necessary to compute (6.21). The construction of the Hessian will therefore only involve a small amount of c.p.u. time and hardly any extra programming effort. The expected adaptation loss equals

$$E(L) = \sum_{i=1}^{n} \left(k_2 \int_{z_i}^{x_i} (x_i - x)^2 h(x)\,dx + k_1 \int_{x_i}^{z_{i+1}} (x - x_i)^2 h(x)\,dx \right) \tag{6.24}$$

$$= \sigma^2 \sum_{i=1}^{n} \left(k_2 \int_{z_i^*}^{v_i} (v_i - v)^2 f(v)\,dv + k_1 \int_{v_i}^{z_{i+1}^*} (v - v_i)^2 f(v)\,dv \right). \tag{6.25}$$

If the size pattern is selected optimally we have after substitution of (6.21) into (6.25),

$$E(L) = \sigma^2 \sum_{i=1}^{n} \left(k_2 \int_{z_i^*}^{v_i} v^2 f(v)\,dv + k_1 \int_{v_i}^{z_{i+1}^*} v^2 f(v)\,dv \right.$$

$$\left. - v_i^2 \left(k_2 \int_{z_i^*}^{v_i} f(v)\,dv + k_1 \int_{v_i}^{z_{i+1}^*} f(v)\,dv \right) \right) \tag{6.26}$$

which reduces, in the special case that $k = k_1 = k_2$, to

$$E(L) = k(\sigma^2 - \sigma_s^2) \tag{6.27}$$

where

$$\sigma_s^2 = \sum_{i=1}^{n} p_i(x_i - \mu_s)^2 \tag{6.28}$$

and

$$\mu_s = \sum_{i=1}^{n} p_i x_i. \tag{6.29}$$

As will be clear, the optimal sizes are located symmetrically around the mean of $h(x)$ if $k_1 = k_2$, therefore $\mu_s = \mu$. The optimal transformed size patterns for $k = k_1 = k_2 = 1$ are listed in Table A.2 of Appendix A for n ranging from 1 to 25. Also the expected adaptation loss of each transformed size pattern is inserted in this table. Multiplying these values by σ^2 we obtain the expected adaptation losses of the nontransformed size patterns.

6.5 THE LINEAR LOSS FUNCTION IN COMBINATION WITH SHIFT RULE T

By virtue of (4.25) we have,

$$k_2(F(v_i) - F(z_i^*)) = k_1(F(z_{i+1}^*) - F(v_i)) \qquad i = 1, \ldots, n \tag{6.30}$$

which gives,

$$v_{i+1} = ((k_1 + k_2) F^{-1} \left(\frac{k_2}{k_1} \{F(v_i) - F(z_i^*)\} + F(v_i) \right)$$
$$- k_1 v_i)/k_2 \qquad i = 1, \ldots, n - 1. \tag{6.31}$$

The optimal transformed size pattern might be calculated using the method of interval bisection. To evaluate the inverse normal distribution function, F^{-1}, the numerical approximation as given by Abramowitz (1972) can be used. Putting $k = k_1 = k_2$, we conclude, since the normal density is symmetrical around its mean,

$$v_i = - v_{n+1-i} \qquad i = 1, \ldots, n. \tag{6.32}$$

If n is odd v_1 must be chosen in such a way that the $((n + 1)/2)$ th size equals zero and if n is even v_1 must be selected such that the absolute value of the $(n/2)$ th size equals the value of the $(n/2 + 1)$th size. Hence, when using the method of interval bisection we never have to calculate more than $(n/2 + 1)$ sizes at each iteration. The expected adaptation loss is,

$$E(L) = \sigma \sum_{i=1}^{n} \left(k_2 \int_{z_i^*}^{v_i} (v_i - v) f(v) dv + k_1 \int_{v_i}^{z_{i+1}^*} (v - v_i) f(v) dv \right) \tag{6.33}$$

which equals,

$$\sigma \sum_{i=1}^{n} (k_1 + k_2)(f(v_i) - f(z_i^*)) \tag{6.34}$$

if the sizes are chosen optimally. The optimal transformed size patterns for $k = k_1 = k_2 = 1$ and n ranging from 1 to 25 are listed in Table A.3 of Appendix A. In Fig. 6.4 the expected adaptation loss is plotted as a function of n for the same values of the parameters. As can be seen, the expected adaptation loss is approximately a hyperbolic function of n. For small values of n, the reduction of E(L) is large if an extra size is added. For larger values of n this reduction, however, decreases sharply. For instance, if n is larger than 25, the percentage reduction of E(L) is less than 4% which generally will hardly compensate the increase of the production and distribution costs caused by the production of an extra size.

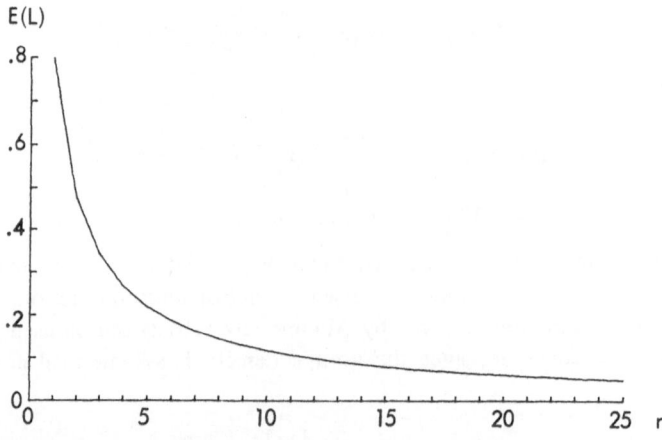

Figure 6.4. The Expected Adaptation Loss as a Function of n

7

THE RECURSION FORMULAE FOR THE LOG NORMAL DISTRIBUTION

7.1 INTRODUCTION

In this chapter we shall substitute the log normal distribution into the recursion formulae derived in Chap. 4. Although the log normal distribution is somewhat less easy to handle than the normal distribution, the methods that are used to reduce the recursion formulae to a more simple form, so that they easily can be calculated, are similar to those of the previous chapter. In this sense, the chapter thus contains no extensions of the theory developed so far. The decision to pay nevertheless some attention to the log normal distribution, however, was mainly taken on the basis of practical considerations. The distribution has namely a rather large field of application since it approximates very well the distribution of demand of a wide variety of different product properties. Moreover, the log normal distribution is often more suited to describe the demand for a product property because it is not defined for negative demands.

From the practical point of view, the log distribution is therefore at least as important as the other three distributions, which is the reason that we discuss it here.

7.2 THE PROBABILITY DENSITY FUNCTION

The log normal distribution has the density function,

$$h(x) = \frac{1}{\sigma\sqrt{2\pi}x} e^{-((\ln x - \mu)/\sigma)^2/2} \quad x > 0,$$

$$= 0 \qquad\qquad\qquad \text{elsewhere,} \qquad\qquad (7.1)$$

where $\ln x$ is the natural logarithm of x. Furthermore, x_0 and x_{n+1} are respectively equal to zero and infinity. If the variable x has the log normal density then the variable v,

$$v = (\ln x - \mu)/\sigma \qquad\qquad (7.2)$$

has the standard normal density. Defining the transformed sizes as,

$$v_i = (\ln x_i - \mu)/\sigma \quad i = 1, \ldots, n+1,$$

$$v_0 = -\infty \qquad\qquad (7.3)$$

it follows that the probability of a required size being smaller than x_i equals

$$H(x_i) = F(v_i). \qquad\qquad (7.4)$$

The mean and the variance of the log normal distribution are denoted respectively by $E(x)$ and $V(x)$. The symbols μ and σ^2 remain reserved for the mean and variance of the normal distribution.

$$E(x) = e^{\mu + \sigma^2/2}, \qquad\qquad (7.5)$$

$$V(x) = e^{2\mu + \sigma^2} (e^{\sigma^2} - 1). \qquad\qquad (7.6)$$

Both formulae can be derived using the result B.48 in Appendix B.

7.3 THE NONLINEAR LOSS FUNCTION IN COMBINATION WITH SHIFT RULE O

The recursion formulae for this case are given by (4.28). Note that the optimal solution depends on the values of the three parameters μ, σ and α and on the factor k/B. Applying (7.2) and (7.3) to (4.28) a set of recursion formulae is obtained in which only the factors $\alpha\beta$, $\alpha\mu$ and k/B appear. After some rearranging we have

$$\alpha\sigma(F(v_i) - F(v_{i-1})) = (e^{\alpha\sigma(v_{i+1} - v_i)} - 1)f(v_i) \quad i = 1, \ldots, n-1, \qquad (7.7)$$

$$(k/B)\alpha\sigma(F(v_n) - F(v_{n-1})) = e^{-\alpha(\sigma v_n + \mu)} f(v_n). \qquad\qquad (7.8)$$

As a result of the transformation and substitutions the log normal integral and function evaluations reduce to standard normal integral and function evaluations. The set of equations (7.7) and (7.8) can be solved with the method of interval bisection since

$$v_{i+1} = \frac{1}{\alpha\sigma}\ln\left(1 + \frac{\alpha\sigma(F(v_i) - F(v_{i-1}))}{f(v_i)}\right) + v_i. \tag{7.9}$$

After the determination of the optimal transformed size pattern, the sizes $x_1, \ldots, x_n$ are calculated according to

$$x_i = e^{\sigma v_i + \mu} \qquad i = 1, \ldots, n. \tag{7.10}$$

The expected adaptation loss equals

$$E(L) = k \sum_{i=1}^{n} x_i^{\alpha} p_i - k \int_0^{x_n} x^{\alpha} h(x)\,dx + B \int_{x_n}^{\infty} h(x)\,dx. \tag{7.11}$$

We call the integral in the second term of (7.11) a partial expectation of x^{α}. By virtue of (B.48) this partial expectation can be reduced to

$$\int_0^{x_n} x^{\alpha} h(x)\,dx = e^{\alpha\mu + (\alpha\sigma)^2/2}\, F(v_n - \alpha\sigma). \tag{7.12}$$

An example of the position of some optimal size patterns, for varying α, is given below. The values of the parameters are

$$\mu = 0, \quad k = 1, \quad n = 5,$$
$$\sigma = 1, \quad B = 100.$$

Table 7.1. Some Optimal Size Patterns for Varying α (Nonlinear Loss Function, Shift Rule O)

α	E(L)	Coverage	x_1	x_2	x_3	x_4	x_5
0.1	0.056	0.999	0.573	1.289	2.800	7.648	76.754
0.5	0.346	0.999	0.749	1.696	3.585	8.728	39.334
1.0	1.032	0.999	0.969	2.154	4.263	8.787	22.488
1.5	2.376	0.996	1.126	2.381	4.295	7.502	13.627
2.0	4.532	0.986	1.197	2.367	3.890	5.968	8.933
5.0	21.103	0.842	1.080	1.623	2.046	2.406	2.727

The table shows that the behavior of x_1, x_2, x_3 and x_4 as a function of α is similar to that of the first four sizes in the example given in Sec. 5.3 (Table 5.1). Size x_5, on the other hand, decreases monotonously as a function of α. This, of course, is caused by the fact that—contrary to the homogeneous distribution— the log normal distribution has no fixed upperbound. We have not treated the linear loss function with shift rule O as a separate case. The results for this loss function namely can be easily obtained by substituting $\alpha = 1$ into the formulae derived above.

7.4 THE QUADRATIC LOSS FUNCTION IN COMBINATION WITH SHIFT RULE T

By virtue of (4.15) and using (7.2), (7.3), (7.4) and (7.12) we have,

$$k_2 \left[x_i(F(v_i) - F(z_i^*)) - E(x)(F(v_i - \sigma) - F(z_i^* - \sigma)) \right]$$
$$= - k_1 \left[x_i(F(z_{i+1}^*) - F(v_i)) - E(x)(F(z_{i+1}^* - \sigma) \right.$$
$$\left. - F(v_i - \sigma)) \right] \qquad i = 1, \ldots, n \qquad (7.13)$$

where

$$z_i^* = (\ln z_i - \mu)/\sigma \qquad i = 2, \ldots, n+1,$$
$$z_1^* = -\infty \qquad (7.14)$$

with z_i defined by (4.5) and (4.17).

The method of interval bisection clearly is not suitable to solve (7.13) so that Newton's method is used. The Hessian, with typical element $\partial^2 E(L)/\partial x_i \partial x_j$, equals,

$$a_{ii} = k_2(F(v_i) - F(z_i^*)) + k_1(F(z_{i+1}^*) - F(v_i))$$
$$- \frac{\sqrt{k_1 k_2}}{\sigma(\sqrt{k_1} + \sqrt{k_2})^2} \left(\frac{k_2(x_i - x_{i-1}) f(z_i^*)}{z_i} \right.$$
$$\left. + \frac{k_1(x_{i+1} - x_i) f(z_{i+1}^*)}{z_{i+1}} \right) \qquad i = 1, \ldots, n,$$

$$a_{i,i+1} = a_{i+1,i} = \frac{-k_1 k_2(x_{i+1} - x_i) f(z_{i+1}^*)}{\sigma(\sqrt{k_1} + \sqrt{k_2})^2 z_{i+1}} \qquad i = 1, \ldots, n-1 \qquad (7.15)$$

and all other elements equal zero. The expected adaptation loss is

$$E(L) = \sum_{i=1}^{n} \left(k_2 \int_{z_i}^{x_i} (x_i - x)^2 h(x) dx \right.$$

$$+ k_1 \int_{x_i}^{z_{i+1}} (x - x_i)^2 \, h(x) \, dx \Bigg) \qquad (7.16)$$

In case the size pattern satisfies (7.13) we may reduce this expression to

$$E(L) = E(x^2) \left(\sum_{i=1}^{n} k_2 (F(v_i - 2\sigma) - F(z_i^* - 2\sigma)) \right.$$

$$\left. + k_1 (F(z_{i+1}^* - 2\sigma) - F(v_i - 2\sigma)) \right) - \sum_{i=1}^{n} \{ x_i^2 (k_2 (F(v_i)$$

$$- F(z_i^*)) + k_1 (F(z_{i+1}^*) - F(v_i))) \}. \qquad (7.17)$$

If $k = k_1 = k_2$, (7.17) can be written as,

$$E(L) = k(V(x) - \sigma_s^2) \qquad (7.18)$$

where σ_s^2 is defined by (6.28).

7.5 THE LINEAR LOSS FUNCTION IN COMBINATION WITH SHIFT RULE T

According to (4.26) we have,

$$x_{i+1} = \frac{k_1 + k_2}{k_2} \, e^{\sigma F^{-1} \, [(k_2/k_1)(F(v_i) - F(z_i^*)) + F(v_i)] + \mu}$$

$$- \frac{k_1 x_i}{k_2} \quad i = 1, \ldots, n - 1 \qquad (7.19)$$

where z_i^* ($i = 1, \ldots, n + 1$) is defined by (7.14). Furthermore the size pattern must satisfy the equation

$$k_2 \{ F(v_n) - F(z_n^*) \} = k_1 (1 - F(v_n)). \qquad (7.20)$$

From (7.19) it follows that the method of interval bisection can be used to locate the optimal size pattern. The expected adaptation loss equals,

$$E(L) = \sum_{i=1}^{n} \left(k_2 \int_{z_i}^{x_i} (x_i - x) \, h(x) \, dx + k_1 \int_{x_i}^{z_{i+1}} (x - x_i) \, h(x) \, dx \right) \qquad (7.21)$$

which can be written in terms of integrals of the standard normal distribution as

$$E(L) = k_1 (E(x) - x_n) + (k_1 + k_2) \sum_{i=1}^{n} \left(x_i F(v_i) - z_i F(z_i^*) \right.$$
$$\left. - E(x)(F(v_i - \sigma) - F(z_i^* - \sigma)) \right) . \qquad (7.22)$$

If the size pattern is chosen optimally this leads to

$$E(L) = E(x) \left(k_1 - (k_1 + k_2) \sum_{i=1}^{n} (F(v_i - \sigma) - F(z_i^* - \sigma)) \right) . \qquad (7.23)$$

8 THE RECURSION FORMULAE FOR THE GAMMA DISTRIBUTION

8.1 INTRODUCTION

The gamma distribution is—as the normal and the log normal distributions—often used to describe the demand for a product property. We mention for instance the demand for the thickness of window panes which follows approximately a gamma distribution [AKB, (1976)].

The gamma distribution has two parameters. By changing the scale of measurement on the x-axis, one of these parameters can be eliminated. We have chosen the transformation such that the transformed gamma variable has a density function which equals that of a χ^2 distribution. Note however, that this does not imply that the transformed gamma variable has a χ^2 distribution; the χ^2 distribution namely is only defined for integer values of its parameter, usually denoted by the term "degrees of freedom," while we will allow it to have fractional values too. The main reason for the choice of this particular transformation is that integrals of the χ^2 distribution are extensively tabulated for integer values of its parameter. This implies that eventually optimal size patterns might be calculated—or approximated if the parameter is fractional—with a simple calculator instead of with a computer.

Of special interest, from the mathematical point of view, is the exponential distribution which is a special case of the gamma distribution. This distribution permits, as we will show, an analytical solution of the optimal size pattern if the linear loss function is used. Besides this, we will prove for the specific case that the linear loss function is used in combination with shift rule O, that the expected adaptation loss reaches a minimum at the size pattern satisfying the recursion formulae.

8.2 THE PROBABILITY DENSITY FUNCTION

The probability density function of a gamma distributed variable is,

$$g(x) = \frac{1}{\Gamma(a)b^a} x^{a-1} e^{-x/b} \qquad x > 0, \ a, b > 0,$$

$$= 0 \qquad\qquad\qquad \text{elsewhere}, \qquad\qquad (8.1)$$

where

$$\Gamma(a) = \int_0^\infty t^{a-1} e^{-t} dt. \qquad\qquad (8.2)$$

By virtue of (8.2),

$$\Gamma(a) = (a - 1) \, \Gamma(a - 1) \qquad\qquad (8.3)$$

and, since $\Gamma(1) = 1$, we may write for integer values of a

$$\Gamma(a) = (a - 1)! \qquad\qquad (8.4)$$

Furthermore, according to (4.1) and (4.2), x_0 equals zero and x_{n+1} equals infinity. The mean and variance of a gamma distributed variable are

$$E(x) = ab, \qquad\qquad (8.5)$$
$$V(x) = ab^2. \qquad\qquad (8.6)$$

If x has a gamma distribution with parameters $a = r/2$ and b, where r is a positive integer, then the random variable v,

$$v = 2x/b \qquad\qquad (8.7)$$

has a χ^2 distribution with r degrees of freedom. The probability density function of v is;

$$h(v) = \frac{1}{\Gamma(r/2)2^{r/2}} v^{r/2-1} e^{-v/2} \qquad v > 0, \ r > 0,$$

$$= 0 \qquad\qquad\qquad \text{elsewhere.} \qquad\qquad (8.8)$$

Putting,

$$v_i = 2x_i/b \qquad i = 0, \ldots, n+1 \qquad (8.9)$$

we may compute the probability that a required size is smaller than x_i with the χ^2 distribution, since

$$G(x_i) = H(v_i). \qquad (8.10)$$

When programming a computer to calculate integrals of the function h the series expansion given by Abramowitz (1972) of page 941 can be used. To compute the term $\Gamma(r/2)$, appearing in this expansion, for fractional values of the factor $r/2$ we write,

$$\Gamma(r/2) = (r/2 - 1)(r/2 - 2) \cdots (r/2 - k) \, \Gamma(r/2 - k) \qquad (8.11)$$

where the integer k is chosen such that

$$0 \leqslant r/2 - k \leqslant 1. \qquad (8.12)$$

Next, $\Gamma(r/2 - k)$ can be calculated using a polynomial approximation [Abramowitz, (1972), see 6.1.36 on page 257]. Note that for odd values of r it is more efficient to use $\Gamma(\tfrac{1}{2}) = \sqrt{\pi}$ instead of the polynomial approximation.

8.3 THE NONLINEAR LOSS FUNCTION IN COMBINATION WITH SHIFT RULE O

By virtue of (4.29) we have, using (8.7) and (8.9)

$$v_{i+1} = v_i \left[\frac{\alpha p_i}{v_i h(v_i)} + 1 \right]^{1/\alpha} \qquad i = 1, \ldots, n-1. \qquad (8.13)$$

Furthermore, the optimal transformed size pattern has, according to (4.28), to satisfy the condition

$$p_n = \left[\frac{B}{k\alpha} \left(\frac{2}{b}\right)^\alpha \right] \frac{h(v_n)}{v_n^{\alpha-1}}. \qquad (8.14)$$

From (8.13) we see that the optimal transformed size pattern can be determined with the method of interval bisection. The location of this size pattern depends only on the respective values of the parameters α and r and on the factor between brackets in (8.14). The expected adaptation loss, written in terms of the transformed sizes, equals

$$E(L) = k \left(\frac{b}{2}\right)^{\alpha} \left[\sum_{i=1}^{n} \{v_i^{\alpha} p_i\} - \frac{\Gamma(a+\alpha)2^{\alpha}}{\Gamma(a)} \int_0^{v_n} \frac{v^{a+\alpha-1}}{\Gamma(a+\alpha)2^{a+\alpha}} e^{-v/2} dv\right]$$
$$+ B(1 - H(v_n)). \tag{8.15}$$

As can be seen, the function behind the integral sign in (8.15) is the function h with parameter $2(a + \alpha)$. The integral thus can be evaluated with the numerical procedures mentioned in Sec. 8.2.

8.4 THE LINEAR LOSS FUNCTION IN COMBINATION WITH SHIFT RULE O

Substitution of $\alpha = 1$ into (8.13) and (8.14) gives the recursion formulae for the linear loss function. The optimal size pattern is calculated in the same way as in Sec. 8.3. The expected adaptation loss, however, can be written in a more simple form, namely

$$E(L) = \frac{kb}{2} \left[\sum_{i=1}^{n} \{v_i p_i\} + 2v_n h(v_n)\right] - (abk + B) H(v_n) + B. \tag{8.16}$$

In the rest of this section we will assume that the probability density function of the required sizes is the exponential distribution. This distribution is obtained from the gamma distribution by putting a equal to unity. The integrals of the exponential distribution can be calculated directly so that a transformation to the density function of the χ^2 distribution is not necessary. Instead we use,

$$v = x/b \tag{8.17}$$

to eliminate the parameter b from the density function. This gives

$$h(v) = e^{-v} \quad v > 0,$$
$$= 0 \quad \text{elsewhere.} \tag{8.18}$$

Using the substitutions,

$$v_i = x_i/b \quad i = 0, \ldots, n+1, \tag{8.19}$$

the recursion formulae (4.22) reduce to

$$e^{-v_{i-1}} - e^{-v_i} = (v_{i+1} - v_i) e^{-v_i} \quad i = 1, \ldots, n-1,$$
$$e^{-v_{n-1}} - e^{-v_n} = (B/kb)e^{-v_n}. \tag{8.20}$$

Hence,

$$v_i = v_{i-1} + \ln(1 + (v_{i+1} - v_i)) \quad i = 1, \ldots, n-1, \tag{8.21}$$
$$v_n = v_{n-1} + \ln(1 + d) \tag{8.22}$$

where

$$d = B/kb. \tag{8.23}$$

Next, the function P is defined as

$$P_j(t) = \ln(1 + P_{j-1}(t)) \quad j = 2, 3, \ldots,$$
$$P_1(t) = \ln(1 + t). \tag{8.24}$$

Combining (8.21), (8.22) and (8.24) we obtain the solution of the recursion formulae

$$v_i = \sum_{j=n+1-i}^{n} P_j(d) \quad i = 1, \ldots, n. \tag{8.25}$$

The expected adaptation loss equals

$$E(L) = kb \sum_{i=1}^{n} \left(\int_{v_{i-1}}^{v_i} (v_i - v)e^{-v} \, dv \right) + B \int_{v_n}^{\infty} e^{-v} \, dv \tag{8.26}$$

$$= kb \sum_{i=1}^{n} \left(v_i(e^{-v_{i-1}} - e^{-v_i}) \right) - kb + e^{-v_n}(kb(v_n + 1) + B). \tag{8.27}$$

Selecting the sizes optimally we conclude that

$$E(L) = kbP_n(d) = kbv_1 = kx_1. \tag{8.28}$$

Finally, we shall prove that the expected adaptation loss reaches a minimum if the sizes are chosen according to (8.25). Since there was only one solution of the recursion formulae this minimum is unique.

The necessary condition for the existence of a minimum is that the matrix of second derivatives is positive definite. Consequently, we have to prove that the leading principal minors[1] of the Hessian, A, are all positive.

The nonzero elements of the Hessian, evaluated at the transformed size pat-

[1] The kth order leading principal minor of a square matrix A is the determinant of the $k \times k$ matrix consisting of the first k rows and columns of A.

tern satisfying the recursion formulae, are — apart from the factor kb — equal to,

$$\frac{\partial^2 E(L)}{\partial v_i^2} = 2e^{-v_i} + (v_{i+1} - v_i)e^{-v_i} = e^{-v_{i-1}} + e^{-v_i} \qquad i = 1, \ldots, n-1,$$

$$\frac{\partial^2 E(L)}{\partial v_i \partial v_{i+1}} = \frac{\partial^2 E(L)}{\partial v_{i+1} \partial v_i} = -e^{-v_i} \qquad i = 1, \ldots, n-1,$$

$$\frac{\partial^2 E(L)}{\partial v_n^2} = e^{-v_n} + de^{-v_n} = e^{-v_{n-1}} \qquad (8.29)$$

so that

$$A = \begin{bmatrix} 1 + e^{-v_1} & -e^{-v_1} & & & & \\ -e^{-v_1} & e^{-v_1} + e^{-v_2} & -e^{-v_2} & & & \\ & \ddots & \ddots & \ddots & & \\ & -e^{-v_{k-1}} & e^{-v_{k-1}} + e^{-v_k} & -e^{-v_k} & & \\ & & \ddots & \ddots & \ddots & \\ & & -e^{-v_{n-2}} & e^{-v_{n-2}} + e^{-v_{n-1}} & -e^{-v_{n-1}} \\ & & & -e^{-v_{n-1}} & e^{-v_{n-1}} \end{bmatrix} \qquad (8.30)$$

Denoting the kth order principal minor by D_k, we have

$$D_1 = 1 + e^{-v_1}, \qquad (8.31)$$
$$D_2 = e^{-v_1} + e^{-v_2} + e^{-(v_1 + v_2)}. \qquad (8.32)$$

Suppose now that D_k equals

$$D_k = e^{-\sum_{i=1}^{k} v_i} \sum_{i=0}^{k} e^{v_i} \qquad k = 1, \ldots, n-1. \qquad (8.33)$$

As readily can be seen (8.31) and (8.32) satisfy (8.33). For $k > 2$ we derive from (8.30), expanding the minor D_k with respect to the first two nonzero elements of the kth row

$$D_k = (e^{-v_{k-1}} + e^{-v_k})D_{k-1} - e^{-2v_{k-1}}D_{k-2} \qquad k = 3, \ldots, n-1. \quad (8.34)$$

D_{k-1} and D_{k-2} can be obtained from (8.33). Substituting these into the right-hand side of (8.34) we find, after some rearranging, that the right-hand side equals D_k. This completes the (inductive) proof that (8.33) is valid. Finally, D_n equals

$$D_n = e^{-v_{n-1}} D_{n-1} - e^{-2v_{n-1}} D_{n-2} = e^{-\sum_{i=1}^{n-1} v_i} \qquad (8.35)$$

From (8.31), (8.32), (8.33) and (8.35) it now follows that every leading principal minor is positive. Therefore the Hessian is positive definite, implying that the expected adaptation loss reaches a minimum at the calculated size pattern (8.25).

8.5   THE QUADRATIC LOSS FUNCTION IN COMBINATION WITH SHIFT RULE T

Applying (8.7) and (8.9) to (4.15) and putting $\alpha = 1$ and $\beta = 2$ we obtain the recursion formulae. These have the same form as those derived in Sec. 6.4 [see (6.21)] except for the fact that the function f must be replaced by the function h. Note that z_i^*, appearing in (6.21), now is defined as

$$z_i^* = 2z_i/b \qquad i = 0, \ldots, n+1. \qquad (8.36)$$

For the calculation of the recursion formulae, use can be made of the fact that

$$\int_0^t (v_i - v) h(v) dv = H(t)(v_i - r) + 2th(t) \qquad (8.37)$$

where $h(v)$ is defined by (8.8) and r equals 2a. Applying Newton's method to find the optimal size pattern we use the Hessian which is, after replacement of the function f by the function h, given by (6.23).

Finally, the expected adaptation loss is given by (6.25). Note that besides changing f in h, also the factor $(b/2)^2$ must be used instead of σ^2. To calculate the expected adaptation loss we have to evaluate integrals of the form

$$\int_0^t (v_i - v)^2 h(v) dv. \qquad (8.38)$$

These may be written in terms of the cumulative distribution function H and the density function h as

$$H(t)((v_i - r)^2 + 2r) + 2th(t)(2v_i - r - 2) - 2t^2 h(t). \qquad (8.39)$$

8.6 THE LINEAR LOSS FUNCTION IN COMBINATION WITH SHIFT RULE T

According to (4.25), the recursion formulae can be written as,

$$k_2 \int_{z_i^*}^{v_i} h(v)dv = k_1 \int_{v_i}^{z_{i+1}^*} h(v)dv \qquad i = 1, \ldots, n \qquad (8.40)$$

where z_i^* is defined by (8.36). Although v_{i+1} can be expressed as an explicit function of v_i and v_{i-1} we apply Newton's method to solve (8.40). This because of the fact that the evaluation of the inverse cumulative distribution function, H^{-1}, requires some rather time consuming calculations. The nonzero elements of the Hessian, which can be derived from (8.40), are equal to

$$a_{ii} = (k_1 + k_2) h(v_i) - \frac{k_2^2}{k_1 + k_2} h(z_i^*) - \frac{k_1^2}{k_1 + k_2} h(z_{i+1}^*) \qquad i = 1, \ldots, n,$$

$$a_{i+1,i} = a_{i,i+1} = \frac{-k_1 k_2}{k_1 + k_2} h(z_{i+1}^*) \qquad i = 1, \ldots, n - 1. \qquad (8.41)$$

The formula for the expected adaptation loss is obtained from (6.33) by replacing σ by the factor $(b/2)$ and using the function h instead of f. The expected adaptation loss might be calculated by means of (8.37).

Next we will consider the recursion formulae for the exponential distribution. It will appear that, also when applying shift rule T, this distribution permits an analytical solution of the optimal size pattern. Putting a equal to unity in (8.1) and applying (8.17) and (8.19) to (4.25) we have the recursion formulae

$$k_2 \{e^{-z_i^*} - e^{-v_i}\} = k_1 \{e^{-v_i} - e^{-z_{i+1}^*}\} \qquad i = 1, \ldots, n \qquad (8.42)$$

where z_i^* equals

$$z_i^* = z_i/b = (k_2 v_i + k_1 v_{i-1})/(k_1 + k_2) \qquad i = 2, \ldots, n + 1,$$

$$z_1^* = 0. \qquad (8.43)$$

Putting the factor $k_1/(k_1 + k_2)$ equal to γ and using (8.43) we can write (8.42) as

$$(1 - \gamma) e^{\gamma(v_i - v_{i-1})} = 1 - \gamma e^{-(1-\gamma)(v_{i+1} - v_i)} \qquad i = 2, \ldots, n, \qquad (8.44)$$

$$(1 - \gamma) e^{v_1} = 1 - \gamma e^{-(1-\gamma)(v_2 - v_1)}. \qquad (8.45)$$

Therefore

$$v_i - v_{i-1} = \frac{1}{\gamma} \ln \left(\frac{1 - \gamma e^{-(1-\gamma)(v_{i+1} - v_i)}}{1 - \gamma} \right) \qquad i = 2, \ldots, n. \qquad (8.46)$$

Because $v_{n+1} - v_n$ equals infinity, we have

$$v_n - v_{n-1} = \frac{1}{\gamma} \ln \left(\frac{1}{1 - \gamma} \right). \qquad (8.47)$$

Substituting this result into (8.46) we obtain $v_{n-1} - v_{n-2}$. Repeating this process we find finally v_1 from (8.45).

In the special case that $k_1 = k_2$ ($\gamma = \frac{1}{2}$) we can express each of the sizes as a function of n. First we put

$$u_i = e^{(v_i - v_{i-1})/2} \qquad i = 2, \ldots, n. \qquad (8.48)$$

By virtue of (8.44) it follows that,

$$u_i = 2 - 1/u_{i+1} \qquad i = 2, \ldots, n. \qquad (8.49)$$

The solution of the difference equation (8.49) is

$$u_i = (n - i + 2)/(n - i + 1) \qquad i = 2, \ldots, n. \qquad (8.50)$$

Substituting (8.48) into (8.46) we obtain

$$\begin{aligned} v_i &= v_{i-1} + 2 \ln (2 - e^{-(v_{i+1} - v_i)/2}) \\ &= v_{i-1} + 2 \ln (u_i) \qquad\qquad i = 2, \ldots, n. \end{aligned} \qquad (8.51)$$

Hence, combining (8.45), (8.50) and (8.51),

$$v_i = \ln \{n(n + 1)/(n + 1 - i)^2\} \qquad i = 1, \ldots, n. \qquad (8.52)$$

From (8.52) we can calculate the probabilities of the service areas. These probabilities form a decreasing arithmetical series with common difference $2/(n(n + 1))$,

$$p_i = 2(n + 1 - i)/n(n + 1). \qquad (8.53)$$

Finally, the expected adaptation loss can be computed using (6.33) by replacing σ and $f(v)$ respectively by b and e^{-v}. Note that the function z_i^* in (6.33) is defined by (8.43). As readily can be verified, if $k = k_1 = k_2$ and the size pattern is optimally chosen, $E(L)$ equals $kbv_1 = kb \ln (1 + 1/n)$. This expression approaches kb/n for large n.

9 THE STANDARDIZATION OF THE POWER OF ELECTRO MOTORS

9.1 INTRODUCTION

In the Netherlands the standardized powers of totally enclosed fan-cooled three-phase squirrel-cage motors with 12.5, 16.66, 25 or 50 revolutions per second, are given by the Nen 3321 standard. Because the motors with 25 r/s are used most often we have chosen this type of motor as an example to illustrate the process of devising an optimal size pattern. We will begin by calculating some optimal size patterns based on the purchase price of 25 r/s motors. Later on, the cost of the power consumed by the motor shall also be taken into consideration.

9.2 THE DEMAND FOR — AND THE PRICE OF — THE 25 r/s MOTOR

The 25 r/s motor is obtainable in 27 sizes, ranging from .06 kW to 132 kW. In total, about 130,000 of these motors are sold yearly in the Netherlands. A frequency table of the estimated total yearly sales (in units) per (standardized) size was kindly put at our disposal by the management of a Dutch firm producing electro motors. Since the sales figures of the two smallest motors (i.e. .06 kW

82

and .09 kW) were not included in this table, we shall consider in the sequel the remaining 25 sizes of the Nen standard only. A graph of the figures showed that a log normal distribution should give a reasonably good fit. In order to estimate the parameters of the density function it was assumed that the users of electro motors apply shift rule O. The major justification for this assumption is the fact that overloading of a motor—which might occur if a motor with a smaller power than the required power is bought—considerably reduces its life time. The estimation was carried out by means of the maximum likelihood method. The model that was used is discussed in the next chapter. The estimated values of the parameters of the log normal density [see (7.1)] are,

$$\hat{\mu} = .178 \quad \text{and} \quad \hat{\sigma} = 1.48 \qquad (9.1)$$

so that according to (7.5) the average required power equals 3.57 kW whereas the standard deviation of the required power equals 10.07 kW. Figure 9.1 displays a part of the density function. The prices of the motors were taken from a list published by the firm mentioned above. It was found that the relationship between the price, π, of a motor and its power, x, is almost exactly linear. Using least squares regression we obtained,

$$\pi(x) = 105.54x + 154.27 \qquad (r = .9999) \qquad (9.2)$$

where π is expressed in glds and x in kW.

Because no information was available about the profit margins of the various

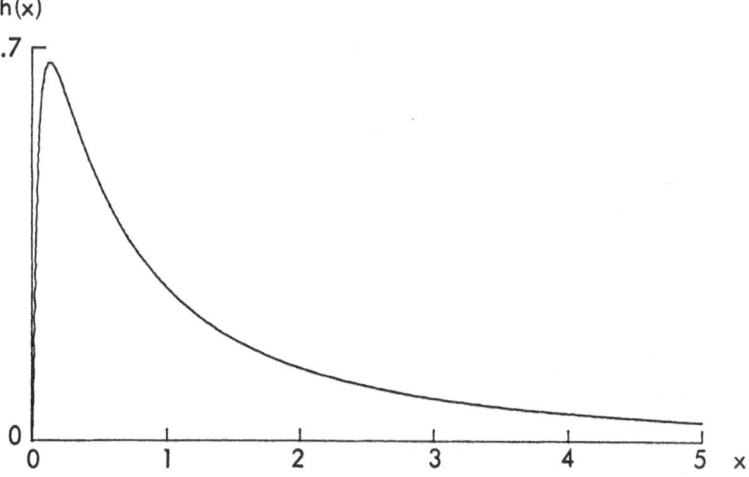

Figure 9.1. The Distribution of the Required Power over the Interval 0 kW—5 kW

types of motors, it was assumed that the price of a motor equals its production costs. Note, that in case the profit margin is a fixed percentage of the production costs of the motor this assumption has no influence on the calculation of an optimal size pattern based on the price of electro motors.

9.3 SOME OPTICAL SIZE PATTERNS BASED ON THE PRICE OF THE 25 r/s MOTORS

By making use of the model discussed in Sec. 7.3 with α equal to 1 and k equal to 105.54 we have calculated the optimal size pattern with its associated expected adaptation loss for n = 25. To facilitate comparisons between the calculated size pattern and the size pattern of the Nen standard the largest size of the first size pattern was put equal to 132 too. The loss of not being served was thus not taken into consideration and, consequently, relation (7.8) was not used. The optimal size pattern and the size pattern of the Nen standard are together with the probabilities of the service areas listed in Table 9.1. As can be seen, the expected adaptation loss of the optimal size pattern equals 47.01 which is about 20% less than that of the size pattern of the Nen standard. The average price that is paid for a motor can be written as

$$\bar{\pi} = \sum_{i=1}^{n} \int_{x_{i-1}}^{x_i} \pi(x_i) h(x) \, dx$$

$$= \sum_{i=1}^{n} \int_{x_{i-1}}^{x_i} \{\pi(x_i) - \pi(x)\} h(x) \, dx + \int_{0}^{x_n} \pi(x) h(x) \, dx. \qquad (9.3)$$

Substituting (9.2) into the last term of (9.3) we obtain

$$\int_{0}^{x_n} \pi(x) h(x) \, dx = 105.54 \int_{0}^{x_n} x h(x) \, dx$$

$$+ 154.27 \int_{0}^{x_n} h(x) \, dx = 514.33. \qquad (9.4)$$

Furthermore, because the first expression of (9.3) equals the expected adaptation loss we have

$$\bar{\pi} = E(L) + 514.33. \qquad (9.5)$$

The use of the Nen standard leads to an average price of 58.47 + 514.33 = 572.80 while when using the optimal size pattern we have $\bar{\pi} = 47.01 + 514.33$

Table 9.1. The Optimal Size Pattern and the Size Pattern of the Nen Standard

Size No.	Optimal Sizes	Prob. Serv. Areas	Nen Standard	Prob. Serv. Areas
1	.27	.156	.12	.060
2	.53	.133	.18	.040
3	.83	.113	.25	.045
4	1.19	.096	.37	.069
5	1.61	.081	.55	.086
6	2.11	.069	.75	.076
7	2.69	.059	1.1	.101
8	3.37	.050	1.5	.083
9	4.18	.043	2.2	.099
10	5.13	.036	3	.073
11	6.24	.031	4	.060
12	7.57	.026	5.5	.056
13	9.14	.022	7.5	.044
14	11.03	.018	11	.040
15	13.30	.015	15	.023
16	16.06	.012	18.5	.012
17	19.44	.010	22	.008
18	23.63	.008	30	.010
19	28.91	.006	37	.005
20	35.65	.005	45	.003
21	44.45	.004	55	.002
22	56.23	.003	75	.002
23	72.52	.002	90	.001
24	96.04	.001	110	.001
25	132.00	.001	132	.000
E(L)	47.01		58.47	
Coverage	.999		.999	

= 561.34. The yearly savings, if instead of the Nen standard the optimal size pattern is used are therefore 130000(572.80 - 561.34) which equals about 1.5 million glds or 2% of the total sales.

When calculating an optimal size pattern consisting of less than 25 sizes, we will obviously find an expected adaptation loss larger than 47.01. An interesting question is: at what number of sizes is the expected adaptation loss of an optimal size pattern approximately equal to that of the Nen standard (i.e. 58.47). By calculating the optimal size patterns with their associated expected adaptation losses for several values of n, we found that the expected adaptation loss of

Table 9.2. The Optimal Size Pattern with 20 Sizes

Size No.	Optimal Sizes	Prob. Serv. Areas	Size No.	Optimal Sizes	Prob. Serv. Areas
1	.33	.193	11	10.64	.024
2	.68	.157	12	13.46	.019
3	1.10	.128	13	17.05	.015
4	1.62	.104	14	21.69	.011
5	2.26	.085	15	27.82	.008
6	3.04	.069	16	36.09	.006
7	4.00	.057	17	47.62	.004
8	5.17	.046	18	64.32	.003
9	6.62	.037	19	89.87	.002
10	8.41	.030	20	132.00	.001
E(L)	59.86				
Coverage	.999				

a size pattern with 20 sizes, which is listed in Table 9.2, approaches this amount closest. Hence, when using a more economical size pattern, the number of sizes can be reduced by 20%, while keeping the expected adaptation loss approximately equal to that of the Nen standard. Note that the calculation above is based on the assumption that the price (= production costs) of a motor, as a function of its power, is still given by (9.2). Since in general the production costs per unit will decrease if n decreases, the calculated reduction of 20% is in fact a lower bound.

Next, we will pay some attention to the determination of the optimal value of n by making use of the model (4.53). In the first place, we need the expected adaptation loss as a function of n. For this purpose, the expected adaptation losses of some optimal size patterns have been calculated for n ranging from 10 to 40. These are presented in Fig. 9.2. The curve drawn in the figure is fitted with the least squares method and it has the equation

$$E(L)_n = 1635.17 \, n^{-1.10}. \tag{9.6}$$

Substituting (9.6) into (9.5) we have

$$\bar{\pi} = 1635.17 \, n^{-1.10} + 514.33. \tag{9.7}$$

Note that (9.7) is in fact only valid for n = 25 since the influence of a changing value of n on the production costs is not yet incorporated in the average price. In the model we use, this influence is measured by the value of ϵ. Taking into

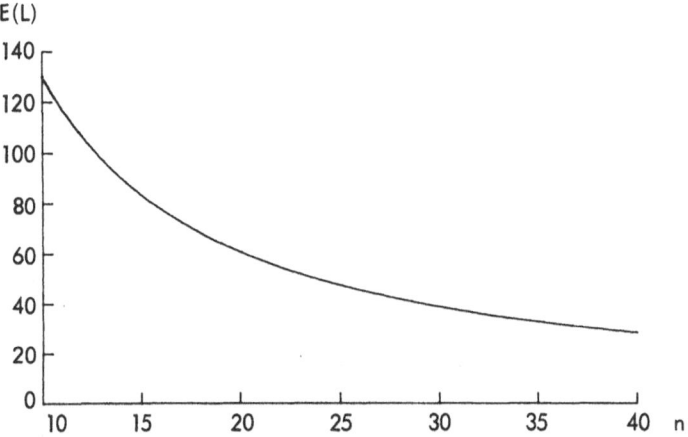

Figure 9.2. The Expected Adaptation Loss as a Function of n

consideration the data supplied by the firm and Table 4.1, it seems reasonable to suppose that the value of ϵ lies—for the production of electro motors—somewhere in the interval (.1, .2). Instead of making an attempt to determine ϵ more precisely—which would require a thorough investigation of the present and eventually future production techniques of electro motors—we will calculate the optimal value of n for three different values of ϵ, namely .1, .15 and .2. In Table 9.3 the average price, corrected for the change in the production costs,

$$\bar{\pi}(n) = (1635.17\, n^{-1.10} + 514.33)(n/25)^{\epsilon} \qquad (9.8)$$

is listed as a function of n for the three values of ϵ. Graphically these functions are presented in Fig. 9.3.

Table 9.3. The Average Price as a Function of n and ϵ

n	$\epsilon = .10$	$\epsilon = .15$	$\epsilon = .20$
10	587.81	561.49	536.34
15	567.73	553.41	539.45
20	562.24	556.00	549.83
25	561.74	561.74	561.74
30	563.30	568.46	573.66
n^{opt}	23	15	11
$\bar{\pi}(n^{opt})$	561.59	553.41	535.70

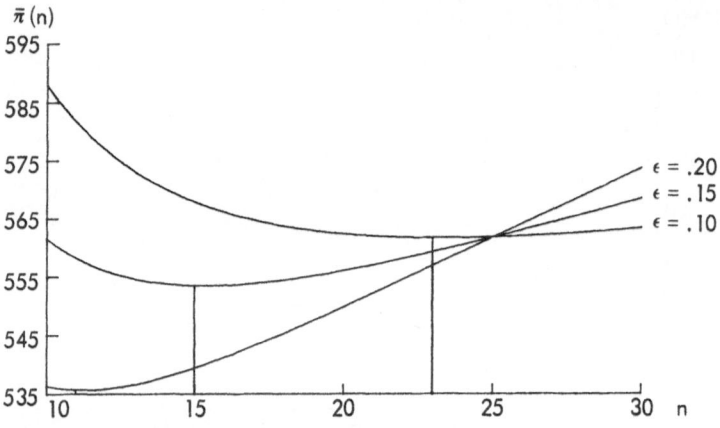

Figure 9.3.   The Average Price as a Function of n and ϵ

As can be seen, if ϵ equals .1, the optimal size pattern consists of 23 sizes. Since this number differs not much from the number of sizes in the present situation, the savings that can be realized are small in this case. If ϵ equals .15, the table shows that the optimal number of sizes equals 15. The savings are then on the average 8.33 glds per motor or expressed as a percentage of the total sales, 1.5% Finally, if ϵ equals .2, the production costs can be reduced by about 4.6% by selecting 11 instead of 25 sizes.

Concluding we remark that—when taking only the price of the motors into account—savings of about 2% of the total sales (in glds) can be realized by choosing the 25 sizes optimally. In case ϵ has a value in the neighborhood of .1, this percentage can hardly be increased by choosing also the optimal number of sizes. If on the other hand ϵ lies between .15 and .2 we may increase the savings to somewhere between the 3.4% and 6.5% by selecting both n and the size pattern optimally. However, all the figures above were calculated under the implicit assumption that the consumption of energy of the motors is independent of the number of sizes and the size pattern chosen. In the next section we will drop this assumption and construct a somewhat more realistic model in which also the cost of energy consumption is incorporated.

9.4   SOME OPTIMAL SIZE PATTERNS BASED ON THE PRICE AND THE ENERGY CONSUMPTION OF 25 r/s MOTORS

The consumption of current of an electro motor is influenced by the following factors:

1. the nominal power of the motor,
2. the load.

If we assume the load to be a constant, say b, times the power of the motor, then the consumption of current can be calculated for some given values of b using a table of efficiency figures. The efficiency of a motor is defined as:

$$e = \frac{\text{Output power}}{\text{Input power}} = \frac{(b)(\text{power of motor})}{\text{Input power}} \qquad b \leqslant 1. \qquad (9.9)$$

For fixed b, the efficiency of a motor is an increasing function of its power. In other words, the larger the capacity of a motor, the less is the use of input power per kW output power. However, if we fix the power of the motor, then the efficiency is maximal for a load equal to this power; if the load decreases then also the efficiency decreases. From the table of efficiency figures, the following relationship was obtained between the power consumption, y, of a motor with power x and load z. The three variables x, y and z are measured in kW.

$$y(z,x) = -.027z \ln x + .020x + 1.164z + .136 \qquad z \leqslant x. \qquad (9.10)$$

An interesting point to note is that – in general – the use of a motor with a power larger than the maximum load is profitable in terms of energy consumption. This is caused by the fact that in most cases the efficiency increase caused by a power increase of the motor is larger than the efficiency decrease caused by the under-loading of the motor. Putting the derivative of (9.10) with respect to x equal to zero, it appears that a motor with power

$$x = \frac{.027}{.020} z = 1.35 z \qquad (9.11)$$

is – in terms of energy consumption – optimal for a load z. To calculate the average hourly consumption of energy during the life of the motor we assumed that the load z can be described by a probability density function specified on the interval $(0, x)$. For this density function we have taken

$$\zeta_x(z) = (t + 1)z^t/x^{t+1} \qquad 0 \leqslant z \leqslant x. \qquad (9.12)$$

Putting $t = 0$, the distribution is the homogeneous one while for t equal to infinity the load equals the power with probability one. For arbitrary t, the average energy consumption is

$$\int_0^x y(z,x)\zeta_x(z)\,dz = -.027\frac{t+1}{t+2}x \ln x + .020\,x$$

$$+ 1.164\frac{t+1}{t+2}x + .136. \qquad (9.13)$$

Multiplying (9.13) with the product, say γ, of the number of hours the motor is working during its life and the cost of a kW hour (kWh) we obtain the total cost of the energy consumed by the motor. Adding the purchase price to this result we have the total cost of a motor with power x

$$g(x) = 105.54\, x + 154.27 + \gamma \left(-.027\, \frac{t+1}{t+2}\, x \ln x + .020\, x \right.$$
$$\left. + 1.164\, \frac{t+1}{t+2}\, x + .136 \right). \tag{9.14}$$

If a motor with power x is required whereas a motor with power x_i ($x_i \geqslant x$) is purchased then the total costs are equal to,

$$k(x,x_i) = 105.54\, x_i + 154.27 + \gamma \int_0^x y(z,x_i) \mathcal{S}_x(z) dz$$
$$= 105.54\, x_i + 154.27 + \gamma \left(-.027\, \frac{t+1}{t+2}\, x \ln x_i \right.$$
$$\left. + .020\, x_i + 1.164\, \frac{t+1}{t+2}\, x + .136 \right) \tag{9.15}$$

so that the adaptation loss is

$$k(x,x_i) - g(x) = (105.54 + .020\gamma)(x_i - x)$$
$$- .027\, \frac{t+1}{t+2}\, \gamma\, x(\ln x_i - \ln x) \qquad x \leqslant x_i. \tag{9.16}$$

For the further calculations we have chosen t equal to one and γ equal to 4380. The latter value is based on a lifetime of a motor of 10 years, 12 working hours a day and a cost of 10 cents per kWh. Substituting these values into (9.16), we have

$$k(x,x_i) - g(x) = 193.14(x_i - x) - 78.84\, x(\ln x_i - \ln x) \qquad x \leqslant x_i. \tag{9.17}$$

Since (9.17)–for a given x–is an increasing function of x_i the assumption that the consumers apply shift rule O is still justified. Note that for large values of γ and t, it may occur that the adaptation loss (9.17) is negative for some x smaller than x_i. This would imply that the optimum power of a motor is larger than its maximum load. In that case shift rule T ought to be applied as–within a certain range–no overloading occurs if a motor with a smaller power than the required (= optimum) power is purchased. The expected adaptation loss equals

$$E(L) = \sum_{i=1}^{n} \int_{x_{i-1}}^{x_i} \{k_1(x_i - x) + k_2\, x(\ln x_i - \ln x)\} h(x) dx \tag{9.18}$$

where $k_2 = -.027\ \gamma(t + 1)/(t + 2)$ and $k_1 = .020\ \gamma + 105.54$. Using transformation (7.2), the substitutions (7.3), formula (B.48) and the result,

$$\int_0^{x_i} x \ln x\ h(x) dx = e^{\mu + \sigma^2/2} [(\mu + \sigma^2)\ F(v_i - \sigma) - \sigma\ f(v_i - \sigma)] \quad (9.19)$$

we may reduce (9.18) to

$$E(L) = \sum_{i=1}^n [k_1 x_i p_i + e^{\mu + \sigma^2/2}\ \{(k_2 \ln x_i - k_1 - k_2\ (\mu + \sigma^2))(F(v_i - \sigma)$$

$$- F(v_{i-1} - \sigma)) - k_2 \sigma(f(v_{i-1} - \sigma) - f(v_i - \sigma))\}]. \quad (9.20)$$

Differentiating (9.18) with respect to x_i $(i = 1, \ldots, n - 1)$ and putting the result equal to zero, we have the recursion formulae

$$k_1 \int_{x_{i-1}}^{x_i} h(x) dx + \frac{k_2}{x_i} \int_{x_{i-1}}^{x_i} x\ h(x) dx = k_1 (x_{i+1} - x_i)\ h(x_i)$$

$$+ k_2 x_i\ (\ln x_{i+1} - \ln x_i)\ h(x_i) \quad i = 1, \ldots, n - 1. \quad (9.21)$$

Although x_{i+1} cannot be written as an explicit function of x_i and x_{i-1}, the solution of (9.21) can be found rather fast with the method of interval bisection. For this purpose, we write the set of equations (9.21) as,

$$k_1 x_{i+1}/x_i + k_2 \ln x_{i+1} - r(x_i, x_{i-1}) = 0 \quad i = 1, \ldots, n - 1, \quad (9.22)$$

where

$$r(x_i, x_{i-1}) = \{\sigma k_1 (F(v_i) - F(v_{i-1}))$$

$$+ \sigma k_2\ e^{\mu + \sigma^2/2}\ (F(v_i - \sigma) - F(v_{i-1} - \sigma))/x_i\}/f(v_i) + k_2 \ln x_i + k_1.$$

For given x_i and x_{i-1}, x_{i+1} now can be found with Newton's method. If we denote the value obtained at the jth step by x_{i+1}^j then x_{i+1}^{j+1} equals

$$x_{i+1}^{j+1} = x_{i+1}^j - \{k_1 x_{i+1}^j/x_i + k_2 \ln x_{i+1}^j$$

$$- r(x_i, x_{i-1})\}/\{k_1/x_i + k_2/x_{i+1}^j\}. \quad (9.23)$$

Convergence is in most cases reached in four or five steps if a reasonable starting value, x_{i+1}^1, is chosen. Good results were obtained by putting x_{i+1}^1 equal to 1.5 x_i.

The optimal size pattern with 25 sizes is listed in Table 9.4. The expected adaptation loss of this size pattern is 54.27. This is 13.47 or 20% less than the expected adaptation loss of the size pattern of the Nen standard which equals 67.74.

Table 9.4. The Optimal Size Pattern with 25 Sizes Based on Price and Current
Consumption

Size No.	Optimal Sizes	Prob. Serv. Areas	Size No.	Optimal Sizes	Prob. Serv. Areas
1	.24	.139	13	8.94	.022
2	.48	.130	14	10.82	.019
3	.77	.114	15	13.08	.015
4	1.11	.098	16	15.84	.013
5	1.52	.084	17	19.23	.010
6	2.00	.072	18	23.44	.008
7	2.56	.061	19	28.74	.006
8	3.23	.052	20	35.52	.005
9	4.02	.045	21	44.38	.004
10	4.96	.038	22	56.23	.003
11	6.06	.032	23	72.62	.002
12	7.37	.027	24	96.18	.001
			25	132.00	.001

E(L)   54.27
Coverage   .999

An optimal size pattern with approximately the same expected adaptation loss
as that of the Nen standard would consist of at most 21 sizes. This implies that
the number of sizes can be reduced by 4 or 16% without letting the expected
adaptation loss becoming larger than that in the present situation.

If we wish to determine the optimal number of sizes, we have to divide the
expected adaptation loss in a part that is included in the production costs and a
part that is independent of these costs. Table 9.5 shows both these parts for n

Table 9.5. The Two Parts of the Expected Adaptation Loss as a Function of n

n	$E(L)_{1,n}$: Part of $E(L)$ That Is Included in the Production Costs	$E(L)_{2,n}$: Part of $E(L)$ That Is Independent of the Production Costs	$E(L) = E(L)_{1,n} + E(L)_{2,n}$
10	131.99	31.76	163.75
15	82.47	15.98	98.45
20	59.93	10.12	70.05
25	47.05	7.22	54.27
30	38.72	5.54	44.26

ranging from 10 to 30. By making use of the same model as in the previous section, we may write the average total costs of a motor as,

$$d(n) = C + (514.33 + E(L)_{1,n})(n/25)^\epsilon + E(L)_{2,n} \qquad (9.24)$$

where C is a constant independent of n. Note that in fact (9.24) is valid only if the term $\Sigma p_i x_i$ is not influenced by ϵ for a given n. Although ϵ has in our problem some influence on the position of the optimal size pattern, and thus on $\Sigma p_i x_i$, this influence is that small that it can be neglected.

The relationship between n and $E(L)_{1,n}$, respectively $E(L)_{2,n}$, again can be approximated very well by a function of the form an^b. Using the data in Table 9.5 we obtain

$$d(n) = C + (514.33 + 1709.48\ n^{-1.12})(n/25)^\epsilon + 1212.44\ n^{-1.59}. \qquad (9.25)$$

Table 9.6 gives the values of d(n) minus C for varying n and the three values of ϵ. As can be seen, due to the inclusion of the power consumption in the model, the optimal values of n are —for all the three values of ϵ—larger than those in the previous section. Furthermore, we conclude from the table that the savings that can be realized by choosing an optimal number of sizes are much larger if ϵ equals .2 than if ϵ equals .1 or .15. This, of course, is due to the fact that for the latter two values of ϵ the optimal value of n lies in the neighborhood of 25. In fact, there exists a value of ϵ between .1 and .15 for which n = 25 is the optimal choice. If, on the other hand ϵ equals .2, the savings that arise when choosing an optimal size pattern consisting of 15 sizes are considerable, namely 130000 (582.07-568.06) is about 1.8 million glds due to the choice of the optimal size pattern and 130000(568.06-555.08) is about 1.7 million glds due to the choice of the optimal n. The total savings are then approximately 3.5 million glds per year or expressed as a percentage of the total sales, 4.7%.

Table 9.6. The Average Total Costs (Minus C) as a Function of n and ϵ

n	$\epsilon = .10$	$\epsilon = .15$	$\epsilon = .20$
10	618.78	592.47	567.33
15	583.32	569.02	555.08
20	571.68	565.45	559.29
25	568.06	568.06	568.06
30	567.81	572.96	578.16
n^{opt}	28	20	15
$d(n^{opt})$	567.65	565.45	555.08

9.5 THE POWER FACTOR

A final important point that we have neglected so far is the power factor. The power factor equals the ratio of the output power (in kW's) and the number of kilovoltamperes that is used by the motor. For instance, if a fully loaded three-phase motor of 3 kW has an efficiency of 81% and a power factor of .83, then it requires,

$$\sqrt{3}\ IV = \frac{\text{Output Power}}{\text{Efficiency} \times \text{Power factor}} = \frac{3}{(.81)(.83)} = 4.46 \qquad (9.26)$$

kilovoltamperes (kVA) from the circuit. If the voltage (V) is 380 V then a current (I) of 6.78 A must be generated and transmitted by the conductors. For the given output power of 3 kW, the lower the power factor, the larger must be the size of the generator to generate that power and the greater must be the cross sectional area of the conductor to transmit it. In other words, the greater is the cost of the generation and the transmission of electrical energy [Hughes, (1963)]. The power factor of a motor depends on the power of the motor and its load in the same way as the efficiency. Thus, the power factor increases with the power of a fully loaded motor and decreases if the load becomes smaller for a given power.

Since we did not take into consideration the power factor in the optimizations above, the situation might arise that the number of kVA the power plant has to generate is larger when using the optimal size pattern than when using the size pattern of the Nen standard. This—as already mentioned above—would increase the cost of generation and transmission of energy.

To compare the two size patterns with respect to this aspect, their expected adaptation losses—in terms of kVA's—were calculated. The relationship between the number of kVA's used, the power of the motor and its load is approximately,

$$y = -.044\ z \ln x + .198\ x + 1.214\ z + .351 \qquad z \leqslant x \qquad (9.27)$$

where y is the number of kVA, z is the load in kW and x is the power in kW. Taking again the parameter t of the probability density function of the load equal to unity, we may write the adaptation loss (in kVA) as

$$E(L) = .198\ (x_i - x) - .044\ \frac{2}{3}\ x\ (\ln x_i - \ln x) \qquad x \leqslant x_i \qquad (9.28)$$

if a motor with power x_i is purchased whereas a motor with power x is required. Note that—as opposed to the adaptation loss measured in kW which could assume negative values for some values of the parameters—(9.28) always is positive. The expected adaptation losses of the optimal size patterns with 15, 20, 25 and 28 sizes, calculated with the loss function (9.28), were respectively equal to .135, .098, .076 and .068 kVA. Within the precision of the three decimals these amounts did not differ from the expected adaptation losses (in kVA) of the optimal size patterns that were determined with the loss function (9.17). On the other hand, the expected adaptation loss of the size pattern of the Nen standard was .095 kVA. Since the expected adaptation loss of the optimal size pattern listed in Table 9.4 is equal to .076 kVA we conclude that the latter size pattern is also more economical in terms of kVA's. If we assume that 130000 25 r/s motors are sold in a year and suppose that these motors are running 12 hours a day with a lifetime of 10 years, then the annual reduction in the generation of current when using the optimal size pattern consisting of 25 sizes instead of the size pattern of the Nen standard is approximately 108 million kVAh.

9.6 SOME ADDITIONAL REMARKS

The application given above may serve as an example of how the standardization of a product (property) should be performed. Because only the data from one firm were analyzed, concerning the demand, the prices, the powers, the efficiencies and the power factors of the motors, the results of the analysis must be considered with care. Furthermore, only the most important aspects of the motors were incorporated in the model; for instance no attention was paid to the maintenance cost and the lifetime, both of which might be influenced by the power and the load of the motor. Also, no research was undertaken to determine the probability density function of the load for a motor with a given power and, finally, requirements as to the (maximal) dimensions of the motor were not taken into consideration. However, it should be no big problem to meet these shortcomings by gathering more data and by extending the model somewhat.

10 THE ESTIMATION OF THE PARAMETERS OF A PROBABILITY DENSITY FUNCTION USING A TABLE OF OBSERVED FREQUENCIES

10.1 INTRODUCTION

In the preceding chapters, the probability density function of the required sizes was assumed to be known. In practical situations, however, this will in general not be the case. Before calculating the optimal size pattern, we then first have to specify a density function, on the basis of a sample, and next its parameters must be estimated.

In principle there are two ways of doing this. In the first place we may sample the population of consumers. The required size(s) and the number of times the product is purchased during a specific period are then recorded for every sampled consumer. From the obtained observations a frequency distribution is constructed, which usually will give an indication as to the type of density function that might be used to describe the behavior of the consumers. After the identification of the density function, its parameters are estimated according to the appropriate sampling theory.

Using the described method, the design of the sampling method may give rise to some problems because of the fact that the buyers of the product often will not be registered. On the other hand, an advantage of the method is the relative simplicity of the estimation procedure.

96

In the second place, it is often very well possible to determine the density function from the sales figures in the past. The sampling units are then—instead of individual consumers—the firms which sell the product. Clearly, this leads to relatively low sampling costs. On the other hand, as most products which are to be standardized are already sold in a certain number of sizes, the use of sales figures usually implies that no exact observations (of the required sizes) are available but only a frequency table of the sales per size. For the specific case that shift rule O is employed, we will show in Sec. 10.2, that this has the effect that application of some commonly used estimators of the parameters of a density function is not possible. In Sec. 10.3 a model, based on the principle of maximum likelihood, will be discussed which we propose to use in this situation.

10.2 SOME ESTIMATION PROCEDURES

To illustrate the various estimation methods, let us suppose that the following frequency table is obtained from the sales figures of a particular product (see Table 10.1). We further assume that in buying the product, the consumers apply shift rule O. The frequency distribution is plotted in Fig. 10.1. This figure indicates that a normal probability density function may give a good fit to the data. The normal distribution is characterized by the parameters μ and σ. The following estimators are commonly used to estimate μ and σ from a frequency table.

$$\hat{\mu} = \sum_{i=1}^{n} f_i m_i / \sum_{i=1}^{n} f_i, \quad \hat{\sigma} = \sqrt{\sum_{i=1}^{n} f_i m_i^2 / \sum_{i=1}^{n} f_i - \hat{\mu}^2}. \tag{10.1}$$

In (10.1), f_i is the number of times a unit with size x_i is purchased and m_i is the middle of the ith service area.

Application of (10.1), however, is not feasible because of the following two reasons. First, it appears from Table 10.1 that the service area of the first size is

Table 10.1. The Frequency Table of the Sales per Size

No	Size	Service Area	Frequency
1	41	$\leqslant 41$	53
2	43	$41 - \leqslant 43$	145
3	45	$43 - \leqslant 45$	267
4	47	$45 - \leqslant 47$	258
5	49	$47 - \leqslant 49$	113

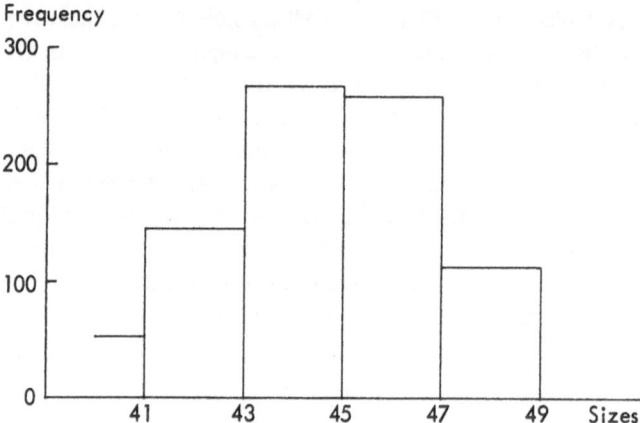

Figure 10.1.   The Frequency Distribution of the Sales Per Size

half open so that the middle of this service area is not defined. A more serious problem, however, is that the frequency, f_{n+1}, of the upper class is missing, which is caused by the use of sales figures. Usually these do not contain information about the number of consumers which could not be served. The effect of the missing frequency is that esimates according to (10.1) will be biased.

Another way of estimating the parameters is to use probability paper of the distribution in question. Before this method can be applied we first have to estimate the frequency in the upper class, which gives rise to problems if the number of sizes is small. Furthermore, a practical disadvantage of the method is the limited applicability because probability paper is only obtainable for a few distributions.

10.3 MAXIMUM LIKELIHOOD ESTIMATION

The method of estimating the parameters of a probability density function which is able to cope with the problems mentioned above is based on the principle of maximum likelihood. Using this principle, the estimators of the parameters are chosen such that they maximize the likelihood function.[1]

---

[1] The form of the likelihood function slightly differs from those used by some writers on the subject, such as Cohen (1950) and Hausman and Wise (1977). The reason is that both these writers assume the individual observations of the truncated sample to be known while we have only a frequency table at our disposal.

$$l(f,\theta) = K \prod_{i=1}^{n} p_i^{f_i} / p^N \qquad (10.2)$$

where

$$p_i = p_i(\theta), \sum_{i=1}^{n} f_i = N \text{ and } \sum_{i=1}^{n} p_i = p \text{ (the coverage)}.$$

The vector θ in the likelihood function denotes the vector of parameters, $(\theta_1, \theta_2, \ldots, \theta_k)$, to be estimated. Furthermore, K is a constant, independent of $\theta_1, \theta_2, \ldots, \theta_k$. The probability that a consumer purchases the ith size is, as usual, denoted by p_i $(i = 1, \ldots, n)$ where p_i is a function of the unknown parameters. Instead of maximizing l we also can maximize ln l because these two functions attain their maximum at the same vector θ. Taking the logarithm of l we have

$$\ln l = \sum_{i=1}^{n} \{f_i \ln p_i\} - N \ln p + \ln K$$

$$= \sum_{i=1}^{n} f_i \ln (p_i/p) + \ln K. \qquad (10.3)$$

This function can be maximized by a nonlinear optimization program. We used a program based on the method of Powel.

An important advantage of the use of. maximum likelihood estimators is the large sample properties these estimators possess. In case the sample size is large, the maximum likelihood estimators are approximately distributed by the multivariate normal distribution with means $\theta_1, \theta_2, \ldots, \theta_k$ and with covariance matrix V [Mood and Graybill, (1950)], where,

$$V = R^{-1} \text{ and } r_{ab} = - E\left(\frac{\partial^2}{\partial\theta_a \partial\theta_b} \ln l(f,\theta)\right) \qquad a, b = 1, \ldots, k. \quad (10.4)$$

These properties enable us to calculate estimates of the standard errors of the parameters which are, as we will see in Chap. 11, important when applying sensitivity analysis to the optimal size pattern. To find the elements of the matrix R we proceed as follows. Differention of ln l with respect to θ_a gives,

$$\partial \ln l / \partial \theta_a = \sum_{i=1}^{n} \{f_i p_i^{(a)} / p_i\} - N p^{(a)} / p \qquad (10.5)$$

where

$$p_i^{(a)} = \frac{\partial}{\partial \theta_a} p_i \qquad i = 1, \ldots, n \text{ and } a = 1, \ldots, k,$$

$$p^{(a)} = \frac{\partial}{\partial \theta_a} \sum_{i=1}^{n} p_i = \sum_{i=1}^{n} p_i^{(a)} \qquad a = 1, \ldots, k. \tag{10.6}$$

The expectation of (10.5) with respect to f is zero, which follows from the fact:

$$E f_i = N p_i / p \qquad i = 1, \ldots, n. \tag{10.7}$$

Hence,

$$\Sigma \Sigma \ldots \Sigma \frac{\partial \ln l}{\partial \theta_a} l = 0 \tag{10.8}$$

where the summations are taken over all sets of values of the f's so that $\Sigma f_i = N$. Differentation of the equality (10.8) with respect to θ_b leads to the relation

$$\Sigma \Sigma \ldots \Sigma \left\{ \frac{\partial^2 \ln l}{\partial \theta_a \partial \theta_b} l + \frac{\partial \ln l}{\partial \theta_a} \frac{\partial \ln l}{\partial \theta_b} l \right\} = 0. \tag{10.9}$$

We conclude

$$E \left(\frac{\partial \ln l}{\partial \theta_a} \right) \left(\frac{\partial \ln l}{\partial \theta_b} \right) = - E \left(\frac{\partial^2 \ln l}{\partial \theta_a \partial \theta_b} \right). \tag{10.10}$$

The typical element r_{ab} of the matrix R therefore equals

$$E \left(\frac{\partial \ln l}{\partial \theta_a} \right) \left(\frac{\partial \ln l}{\partial \theta_b} \right) = E \left(\sum_{i=1}^{n} \left\{ \frac{f_i p_i^{(a)}}{p_i} \right\} \right.$$

$$\left. - \frac{N p^{(a)}}{p} \right) \left(\sum_{i=1}^{n} \left\{ \frac{f_i p_i^{(b)}}{p_i} \right\} - \frac{N p^{(b)}}{p} \right). \tag{10.11}$$

Since,

$$E(f_i f_j) = N(N - 1) p_i p_j / p^2 \qquad i \neq j, \; i, j = 1, \ldots, n,$$
$$E(f_i^2) = N(N - 1) p_i^2 / p^2 + N p_i / p \qquad i = 1, \ldots, n \tag{10.12}$$

we have

$$r_{ab} = \frac{N}{p} \left(\sum_{i=1}^{n} \left\{ \frac{p_i^{(a)} p_i^{(b)}}{p_i} \right\} - \frac{p^{(a)} p^{(b)}}{p} \right). \tag{10.13}$$

To illustrate the use of the method suppose that the probability density function is normal with unknown parameters μ $(= \theta_1)$ and σ $(= \theta_2)$. In that case

$$\partial p_i/\partial\mu = p_i^{(1)} = -\,(f(v_i) - f(v_{i-1}))/\sigma \qquad i = 1, \ldots, n,$$
$$\partial p_i/\partial\sigma = p_i^{(2)} = -\,(v_i f(v_i) - v_{i-1} f(v_{i-1}))/\sigma \qquad i = 1, \ldots, n. \qquad (10.14)$$

The function f in (10.14) is the standard normal distribution. The v_i $(i = 1, \ldots, n)$ are the transformed sizes

$$v_i = (x_i - \mu)/\sigma \qquad i = 0, \ldots, n. \qquad (10.15)$$

As is easy to see,

$$p^{(1)} = -\,f(v_n)/\sigma \quad \text{and} \quad p^{(2)} = -\,v_n\, f(v_n)/\sigma. \qquad (10.16)$$

It then follows that the elements of the matrix R are equal to

$$r_{11} = \frac{N}{p\sigma^2}\left[\sum_{i=1}^{n}\left\{\frac{(f(v_i) - f(v_{i-1}))^2}{p_i}\right\} - \frac{f(v_n)^2}{p}\right], \qquad (10.17)$$

$$r_{12} = r_{21} = \frac{N}{p\sigma^2}\left[\sum_{i=1}^{n}\left\{\frac{(f(v_i) - f(v_{i-1}))(v_i f(v_i) - v_{i-1} f(v_{i-1}))}{p_i}\right\}\right.$$
$$\left. - \frac{v_n f(v_n)^2}{p}\right], \qquad (10.18)$$

$$r_{22} = \frac{N}{p\sigma^2}\left[\sum_{i=1}^{n}\left\{\frac{(v_i f(v_i) - v_{i-1} f(v_{i-1}))^2}{p_i}\right\} - \frac{v_n^2 f(v_n)^2}{p}\right]. \qquad (10.19)$$

If the number of sizes approaches infinity and the transformed sizes are equally distributed on the interval $(-t, t)$ the limiting form of the elements of R as t becomes large can be calculated as follows. Taking for instance the element r_{11}, we see that the last term between the brackets of (10.17) approaches zero. Since p approaches unity for large t, (10.17) might be written as,

$$\frac{N}{\sigma^2}\sum_{i=1}^{n}(v_i - v_{i-1})\left\{\frac{(f(v_i) - f(v_{i-1}))^2}{(v_i - v_{i-1})^2}\right\}\bigg/\left\{\frac{F(v_i) - F(v_{i-1})}{v_i - v_{i-1}}\right\} \qquad (10.20)$$

where F is the cumulative distribution function of the standard normal distribution. The summation in (10.20) takes place over all the transformed sizes of the size pattern. Because the number of sizes approaches infinity, the difference between two successive sizes approaches zero. By virtue of the mean value

theorem and the definition of the Riemann integral we then may approximate
(10.20) with,

$$\frac{N}{\sigma^2} \int_{-t}^{t} \frac{\{\partial f(v)/\partial v\}^2}{\partial F(v)/\partial v} \, dv = \frac{N}{\sigma^2} \int_{-t}^{t} v^2 f(v) dv \qquad (10.21)$$

which approaches N/σ^2 for large t.

In the same way we calculate the limiting form of the other elements of R
which gives

$$R = \frac{N}{\sigma^2} \begin{bmatrix} 1 & 0 \\ 0 & 2 \end{bmatrix}. \qquad (10.22)$$

The covariance matrix is equal to

$$V = \begin{bmatrix} \dfrac{\sigma^2}{N} & 0 \\ 0 & \dfrac{\sigma^2}{2N} \end{bmatrix}. \qquad (10.23)$$

As could be expected, this matrix equals the covariance matrix we find in case
the calculations are based upon the separate values of the sample observations
and not upon a frequency distribution.

The information we lose, because the observations are put into a frequency
table, is measured by the difference in the standard errors of the parameters. In
general, the values of these standard errors will increase if the number of classes
in the frequency table decreases.

10.4   RESULTS OF THE ESTIMATION PROCEDURE

The data in Table 10.1 give rise to the following estimates for μ and σ

$$\hat{\mu} = 44.7515, \quad \hat{\sigma} = 2.3823. \qquad (10.24)$$

In fact the frequency distribution was generated from a normal distribution with
mean $\mu = 44.765$ and standard deviation $\sigma = 2.39$, so the found estimates turned
out to be fairly good. The covariance matrix V, calculated using the estimates
(10.24), equals

$$V = \begin{bmatrix} .0100 & .0032 \\ .0032 & .0076 \end{bmatrix}. \qquad (10.25)$$

The respective standard errors of $\hat{\mu}$ and $\hat{\sigma}$ are .1003 and .0871 and the correlation coefficient between $\hat{\mu}$ and $\hat{\sigma}$ is equal to .3678.

Because the distribution of $\hat{\mu}$ and $\hat{\sigma}$ is approximately bivariate normal we can use the covariance matrix to construct a confidence interval, in the form of an ellipse, for the parameters μ and σ. This confidence interval can be, as is already mentioned in Sec. 10.3, useful when applying sensitivity analysis to the optimal size pattern.

The final matter to be taken into consideration in this chapter is the "testing" of the goodness of fit. We used the test criterion,

$$\chi^2 = \sum_{i=1}^{n} \frac{(f_i - N\,p_i/p)^2}{N\,p_i/p} \tag{10.26}$$

which in large samples has approximately a chi-square distribution with $n - 3$ degrees of freedom in case the frequency distribution is generated from the estimated normal density function. The value of χ^2 was found to be .967 which indicates a very good fit between the estimated density and the frequency distribution, if we compare this value with the value 5.99, corresponding to a .05 significance level of the chi-square distribution.

11 FURTHER ANALYSIS OF OPTIMAL SIZE PATTERNS

11.1 INTRODUCTION

In this chapter three subjects are discussed which are related to the choice of an optimal size pattern. In the first place we will consider what consequences—in terms of the conditional expected adaptation losses—this choice may have for the various subpopulations. Here we use a technique which is known as "analysis of variance."

Secondly, when having calculated an optimal size pattern, it is of importance to know how far the parameter values can vary without causing violent changes in the computed expected adaptation loss. Such an investigation is termed a sensitivity analysis; its use is illustrated by two examples in Sec. 11.3.

In the third place we discuss in Sec. 11.4 briefly the question of the uniqueness of a size pattern satisfying the recursion formulae. It will be shown that there may exist more than one solution and that, consequently, the expected adaptation loss reaches in that case one or more subminima, submaxima or points of inflexion. Though no mathematical proofs are given, numerical experimentation indicates that those cases are the exception rather than the rule.

11.2 ANALYSIS OF VARIANCE

11.2.1 Implications of the Choice of an Optimal Size Pattern

Application of the objective of the minimization of the expected adaptation loss leads to size patterns which optimize the position of the consumers of the product as a group. We further will call this group the population.

No attention has yet been paid to the consequences of the choice of an optimal size pattern for individual consumers or for certain groups in the population (subpopulations). As is already pointed out in Chap. 3, some consumers or subpopulations might be seriously injured by this choice. To show what are the circumstances under which this is likely to occur, let us suppose that the required sizes of every consumer over a given period are drawings from a probability distribution with every consumer having its own probability density function. Clearly, the probability density function of the required sizes then is a weighted average of these density functions, the weights being the number of times the product is bought by the different consumers. We now distinguish between the following situations. In the first place, it may occur that the density functions of the individual consumers are all equal to each other. In that case, independent of the weights, the probability density function of the required sizes equals that of the consumers. In this situation it is obvious that the adaptation losses will be divided evenly over the consumers, or more specifically, over a long period of time the average adaptation loss per unit approaches for each consumer the computed expected adaptation loss, $E(L)$.

On the other hand, however, we may have the situation that the density functions of the consumers all differ from each other. It may even very well occur that these density functions are degenerate, which means that the required size of a consumer is always (about) the same. This for instance is the case with the demand of adults for shoe sizes.

If each consumer is characterized by a specific demand, which is the assumption we will adopt in this and the following section, we can split up the population into a number of subpopulations as soon as a size pattern is chosen. The ith $(i = 1, \ldots, n)$ subpopulation then consists of the consumers who purchase the ith size. Having defined the subpopulations we can determine the consequence of the choice of an optimal size pattern for each of the subpopulations by calculating the conditional expected adaptation losses. Of special interest are then the differences between these losses and their deviation from the $E(L)$. A useful measure for the dispersion of the conditional expected adaptation losses around the $E(L)$ is given by the weighted sum of squared differences, the weights being the probabilities of the service areas. We call this measure the variance between the subpopulations. A small value of this variance indicates that the losses are divided about equally over the subpopulations while a large value means the opposite. In the latter case, action might be undertaken to redistribute the losses more evenly, for instance by means of taxes and subsidies if the adaptation loss can be expressed in monetary terms. In addition, we will show how to calculate the variance of the adaptation loss within the overall population and the variances of the adaptation loss within each of the subpopulations. Both of these variances give an indication as to the dispersion of the adaptation losses (per unit) suffered by individual consumers.

11.2.2 The Calculation of the Variances

We will illustrate the calculation of the variances for the specific case that the linear loss function is used in combination with shift rule O. Furthermore we assume x_n to be fixed at a value that large that the terms,

$$\int_{x_0}^{x_n} h(x)\,dx, \qquad \int_{x_0}^{x_n} x h(x)\,dx \quad \text{and} \quad \int_{x_0}^{x_n} x^2 h(x)\,dx,$$

approximately equal respectively unity, $E(x)$ and $E(x^2)$. Consequently, we do not take into consideration the losses of consumers not being served ($B = 0$). We begin by calculating the variance of the expected loss within the total population. This variance is denoted by $V(L)_t$. We have

$$V(L)_t = E(L^2) - E(L)^2. \qquad (11.1)$$

The first term of the right-hand side of (11.1) equals

$$E(L^2) = k^2 \sum_{i=1}^{n} \int_{x_{i-1}}^{x_i} (x_i - x)^2 h(x)\,dx$$

$$= k^2 \left\{ \sum_{i=1}^{n} x_i^2 \int_{x_{i-1}}^{x_i} h(x)\,dx \right.$$

$$\left. - 2 \sum_{i=1}^{n} x_i \int_{x_{i-1}}^{x_i} x h(x)\,dx + \int_{x_0}^{x_n} x^2 h(x)\,dx \right\}. \qquad (11.2)$$

Next, we define q_i $(i = 1, \ldots, n)$ as

$$q_i = \left(\int_{x_{i-1}}^{x_i} xh(x)dx \right) / p_i \quad i = 1, \ldots, n. \tag{11.3}$$

The variable q_i $(i = 1, \ldots, n)$ equals the conditional expectation of the required size of a consumer given that he purchases the ith size. Using (11.3) and the assumptions made we may write (11.2) as

$$E(L^2) = k^2 \left\{ \sum_{i=1}^n p_i x_i^2 - 2 \sum_{i=1}^n p_i q_i x_i + E(x^2) \right\}. \tag{11.4}$$

For the calculation of $V(L)_t$ we further need the expected adaptation loss which equals

$$E(L) = k \sum_{i=1}^n \int_{x_{i-1}}^{x_i} (x_i - x)h(x)dx = k \left\{ \sum_{i=1}^n p_i x_i - E(x) \right\}. \tag{11.5}$$

Since the sum of the probabilities of the service areas approximately equals one we may interpret $\Sigma p_i x_i$ as the expected value of the purchased sizes. This expectation is denoted by μ_s [see also (6.29)], so (11.5) can be written as

$$E(L) = k(\mu_s - E(x)). \tag{11.6}$$

Note that according to (11.6) the expected adaptation loss might be interpreted as k times the difference between the expected value of the (n) purchased sizes and the expected value of the purchased sizes if the size pattern would have consisted of an infinite number of sizes, evenly distributed over the region of demand. Substituting (11.4) and (11.6) into (11.1) we obtain,

$$V(L)_t = k^2 \left\{ \sum_{i=1}^n p_i x_i^2 - 2 \sum_{i=1}^n p_i q_i x_i + E(x^2) - \mu_s^2 + 2 E(x) \mu_s - E(x)^2 \right\}$$

$$= k^2 \left\{ \sigma_s^2 + V(x) - 2 \sum_{i=1}^n p_i (x_i - \mu_s)(q_i - E(x)) \right\} \tag{11.7}$$

where σ_s^2 is defined by (6.28). As can be seen, the last term between the braces of (11.7) equals twice the covariance between the sizes purchased and the conditional expectations of the required sizes. We can reduce (11.7) to a more simple expression if we assume the density function of the required sizes to be normal with mean μ and standard deviation σ. Selecting the sizes optimally it follows from (4.22) and (B.13) that

$$1 - \int_{x_{n-1}}^{x_n} h(x)\,dx = \sum_{i=1}^{n-1} \int_{x_{i-1}}^{x_i} h(x)\,dx = \sum_{i=1}^{n-1} (x_{i+1} - x_i)h(x_i)$$

$$= \sum_{i=1}^{n} \{x_i(h(x_{i-1}) - h(x_i))\} + x_n h(x_n)$$

$$= \sum_{i=1}^{n} \left\{ p_i \frac{x_i(q_i - \mu)}{\sigma^2} \right\} + x_n h(x_n). \tag{11.8}$$

Substitution of this result into (11.7) gives

$$V(L)_t = k^2 \{\sigma_s^2 - \sigma^2 + 2\sigma^2(p_n + x_n h(x_n))\}. \tag{11.9}$$

Numerical experimentation indicates that formula (11.9) gives a good approximation of the variance if x_n is selected larger than $\mu + 3\sigma$.

The variance $V(L)_t$ is a measure for the differences between the adaptation losses (per unit) of individual consumers. A large value of $V(L)_t$ indicates that some consumers suffer a large adaptation loss per unit while others suffer a small adaptation loss per unit. This however is inherent to the concept of standardization; unless an infinite number of sizes is selected it is unavoidable that some consumers are better off than others. More important than the differences between the adaptation losses of the individual consumers are the differences between the expected adaptation losses of the various subpopulations. A measure for these differences is the variance between the subpopulations, $V(L)_b$. To calculate $V(L)_b$ we need the conditional expected losses of the consumers in the subpopulations. These equal

$$E(L)_i = \frac{k \int_{x_{i-1}}^{x_i} (x_i - x)h(x)\,dx}{p_i}$$

$$= k(x_i - q_i) \qquad\qquad i = 1, \ldots, n. \tag{11.10}$$

We have

$$V(L)_b = \sum_{i=1}^{n} p_i(E(L)_i - E(L))^2. \tag{11.11}$$

In the extreme case $V(L)_b$ equals zero which means that the conditional expected adaptation losses of the subpopulations are equal to each other. This situation occurs if the probability density function is the homogeneous distribution and the linear loss function is used. If the probability density function is the normal distribution and the linear loss function is used in combination with shift rule O, $V(L)_b$ turned out to be rather large if the size pattern has a coverage near one and if the sizes are chosen according to the objective of the minimum expected adaptation loss. To illustrate this point, we have inserted into Table 11.1 the optimal size pattern for the parameter values

$$\mu = 10, \quad n = 10,$$
$$\sigma = 1, \quad k = 1.$$

The largest size x_n is fixed at 13.5. As can be seen the conditional expected losses of the subpopulations 3 to 7 are about equal to each other. The conditional expected loss of the 10th subpopulation is relatively large; this value is mainly responsible for the rather large standard deviation between the subpopulations. Though the conditional expected loss of the 10th subpopulation is large we see from Table 11.1 that the total (expected) loss of this subpopulation only equals .015 × 1.00 = .015 or about 6% of the total expected adaptation loss of .245. Trying to reduce the conditional expected loss for this subpopulation by adding an extra size between x_9 and x_{10} seems therefore hardly of interest. We might however consider to fix x_{10} at a lower value, say 13. This of course reduces the coverage of the size pattern but on the other hand, as can be seen from Table 11.2, the conditional expected loss of the 10th subpopulation decreases considerably.

We have inserted into both tables also the conditional variances (and standard deviations) of the losses in the subpopulations. These variances equal

$$V(L)_i = \left(\int_{x_{i-1}}^{x_i} (k(x_i - x) - E(L)_i)^2 h(x) \, dx \right) / p_i \quad i = 1, \ldots n. \quad (11.12)$$

The variances give just as the total variance, $V(L)_t$, information about the differences between the expected adaptation losses of individual consumers and are therefore, when judging the performance of the size pattern, of less importance than $V(L)_b$. Finally, because the variances $V(L)_i$ ($i = 1, \ldots, n$) and $V(L)_b$ are the components of $V(L)_t$ we have

$$V(L)_t = V(L)_b + \sum_{i=1}^{n} p_i \, V(L)_i. \quad (11.13)$$

Table 11.1. The (Conditional) Expected Adaptation Losses and Variances in the Subpopulations and in the Overall Population ($x_n = 13.5$)

Size No.	Optimal Sizes	Probabilities Service Areas	Conditional Expectations of the Required Sizes	Conditional Expected Losses in the Subpopulations	Conditional Variances in the Subpopulations	Conditional Standard Deviations in the Subpopulations
1	8.595	.080	8.142	.453	.158	.397
2	9.133	.113	8.891	.242	.023	.153
3	9.545	.132	9.348	.197	.014	.119
4	9.911	.140	9.731	.180	.011	.105
5	10.264	.139	10.087	.177	.010	.101
6	10.626	.130	10.440	.186	.011	.104
7	11.022	.112	10.813	.209	.013	.114
8	11.497	.086	11.236	.261	.018	.136
9	12.161	.052	11.764	.396	.034	.184
10	13.500	.015	12.497	1.003	.082	.287
Coverage		+ 1.000				

Expected adaptation loss: .245
Variance between subpopulations: .016 (standard deviation: .125)
Mean variance in subpopulations: + .028 (standard deviation: .166)

Total variance: .043 (standard deviation: .208)

Table 11.2. The (Conditional) Expected Adaptation Losses and Variances in the Subpopulations and in the Overall Population ($x_n = 13$)

Size No.	Optimal Sizes	Probabilities Service Areas	Conditional Expectations of the Required Sizes	Conditional Expected Losses in the Subpopulations	Conditional Variances in the Subpopulations	Conditional Standard Deviations in the Subpopulations
1	8.571	.077	8.122	.450	.156	.395
2	9.104	.109	8.865	.239	.023	.152
3	9.510	.127	9.316	.194	.014	.117
4	9.869	.136	9.693	.176	.011	.103
5	10.213	.136	10.041	.172	.010	.099
6	10.562	.129	10.384	.179	.010	.101
7	10.940	.113	10.742	.198	.012	.109
8	11.382	.090	11.143	.240	.016	.126
9	11.970	.059	11.629	.341	.027	.165
10	13.000	.023	12.292	.708	.064	.254

Coverage   + .999

Expected adaptation loss: .236
Variance between subpopulations: .011 (standard deviation: .106)
Mean variance in subpopulations: .026 (standard deviation: .162)
                              + _____
Total variance: .037 (standard deviation: .193)

111

11.3 SENSITIVITY ANALYSIS

In order to evaluate the consequences of an error in the estimation of one of the cost parameters or one of the parameters of the probability density function of the required sizes we may apply a sensitivity analysis to the calculated size pattern. To illustrate the use of such an analysis we will give two examples. In the first example, concerning the homogeneous distribution in combination with the linear loss function, a mathematical expression is derived which measures the effect of a change of the costs parameters k_1 and k_2 (see Sec. 5.5). In general, however, such an expression cannot be derived in which case the sensitivity of the expected loss with respect to changes in the parameter values must be calculated numerically. This is done in the second example where the normal distribution is used with the linear loss function (see Sec. 6.3).

Example 1. Assuming the density function of the required sizes to be the homogeneous distribution and the loss function to be linear in combination with shift rule T the optimal size pattern equals, according to (5.35),

$$x_i = a + \frac{b-a}{n} \left(\frac{k_1}{k_1 + k_2} + i - 1 \right) \qquad i = 1, \dots, n. \qquad (11.14)$$

with

$$E(L) = (b-a)k_1 k_2 / 2n(k_1 + k_2). \qquad (11.15)$$

Suppose now that the cost parameters equal $\delta_1 k_1$ and $\delta_2 k_2$ instead of k_1 and k_2. The expected adaptation loss—which can be computed by making the appropriate substitutions in (4.10)—then equals

$$E(L) = (b-a)\, k_1 k_2\, (\delta_2 k_1 + \delta_1 k_2)/2n(k_1 + k_2)^2. \qquad (11.16)$$

Unless δ_1 equals δ_2 the size pattern (11.14) clearly is not optimal anymore. The optimal size pattern based upon $\delta_1 k_1$ and $\delta_2 k_2$ namely is

$$x_i = a + \frac{b-a}{n} \left(\frac{\delta_1 k_1}{\delta_1 k_1 + \delta_2 k_2} + i - 1 \right) \qquad i = 1, \dots, n \qquad (11.17)$$

with

$$E(L) = (b-a)\delta_1 \delta_2 k_1 k_2 / 2n(\delta_1 k_1 + \delta_2 k_2). \qquad (11.18)$$

The increase of the expected adaptation loss due to the wrong estimation of the cost parameters is

$$\frac{b-a}{2n} k_1 k_2 \left\{ \frac{\delta_2 k_1 + \delta_1 k_2}{(k_1 + k_2)^2} - \frac{\delta_1 \delta_2}{\delta_1 k_1 + \delta_2 k_2} \right\}$$

$$= \frac{(b-a) k_1^2 k_2^2 (\delta_1 - \delta_2)^2}{2n (k_1 + k_2)^2 (\delta_1 k_1 + \delta_2 k_2)}$$

or

$$\frac{k_1 k_2 (\delta_1 - \delta_2)^2}{(k_1 + k_2)^2 \delta_1 \delta_2} \; 100\% \qquad (11.19)$$

of (11.18). Putting $k_2 = \gamma k_1$ and $\delta_2 = \zeta \delta_1$, we may write the percentage increase of the expected adaptation loss as

$$\frac{\gamma}{(1 + \gamma)^2} \frac{(1 - \zeta)^2}{\zeta} \; 100\%. \qquad (11.20)$$

Table 11.3 shows the results of (11.20) for values of γ and ζ ranging between .1 and 10. As can be seen from the table, if ζ is in the interval (.6, 1.67) the percentage increase of the expected adaptation loss is less than 6.67% for all values of γ. This clearly indicates that the sensitivity of the expected adaptation loss with respect to k_1 and k_2 is low.

Example 2. We now turn to the somewhat more complicated case of the testing of the sensitivity of the expected adaptation loss for changes in the parameters, if the probability density function of the required sizes is normal and the loss function is linear. Furthermore we now assume shift rule O to be applied.

Since it is not possible to express the expected adaptation loss as a simple function of the parameters, we have inserted into Table 11.4 some numerical results concerning the increase of the expected loss in case one of the parameters k, B, σ or μ is wrongly estimated. The optimal size pattern has been calculated using the parameter values,

$$k = 1, \quad \mu = 10, \quad n = 10,$$
$$B = 10, \quad \sigma = 1.$$

which gives

x_1	x_2	x_3	x_4	x_5	x_6	x_7	x_8
8.586	9.122	9.532	9.895	10.244	10.601	10.990	11.452

x_9	x_{10}
12.084	13.285

$$(11.21)$$

The expected adaptation loss of this size pattern equals .247.

Table 11.3. The Percentage Increase of E(L) as a function of γ and ζ

γ	ζ								
	.10 10.00	.20 5.00	.30 3.33	.40 2.50	.50 2.00	.60 1.67	.70 1.43	.80 1.25	.90 1.11
.10 10.00	66.94	26.45	13.50	7.44	4.13	2.20	1.06	.41	.09
.20 5.00	112.50	44.44	22.69	12.50	6.94	3.70	1.79	.69	.15
.30 3.33	143.79	56.80	28.99	15.98	8.88	4.73	2.28	.89	.20
.40 2.50	165.31	65.31	33.33	18.37	10.20	5.44	2.62	1.02	.23
.50 2.00	180.00	71.11	36.30	20.00	11.11	5.93	2.86	1.11	.25
.60 1.67	189.84	75.00	38.28	21.09	11.72	6.25	3.01	1.17	.26
.70 1.43	196.19	77.51	39.56	21.80	12.11	6.46	3.11	1.21	.27
.80 1.25	200.00	79.01	40.33	22.22	12.35	6.58	3.17	1.23	.27
.90 1.11	201.94	79.78	40.72	22.44	12.47	6.65	3.21	1.25	.28
1.00 1.00	202.50	80.00	40.83	22.50	12.50	6.67	3.21	1.25	.28

The first part of Table 11.4 gives the results for values of k between .5 and 1.5. If for instance the true value of k equals .5 instead of 1 we see that the expected adaptation loss of size pattern (11.21) equals .126, whereas the expected adaptation loss of the optimal size pattern—computed with k = .5—equals .125. The differences between the expected adaptation losses—expressed as a percentage of the expected adaptation loss of the optimal size pattern—have been inserted into column 4. The values in this column indicate that the sensitivity of the expected adaptation loss with respect to k is low. The sensitivity of the expected adaptation loss with respect to the parameter B is also low, as can be seen from the second part of Table 11.4. The third part of the table contains the results for varying σ. It appears that a wrong estimation of this parameter incurs larger increases of the expected adaptation loss than the previous two parameters. Most sensitive however is the expected adaptation loss with respect to changes in μ. From Part IV of the table we see that a rather small error in the estimation of μ leads to a fairly large increase of E(L). This result is of course also caused by the value of σ. In this example we have chosen σ equal to unity; for larger values of σ the sensitivity of the expected adaptation loss with respect to μ will obviously decrease.

In the example given above, we have rather arbitrarily chosen the bounds within which the parameter values vary. In practice one will choose these bounds such that they contain the "true" parameter value with reasonable certainty. In general, the expected adaptation loss will be less influenced by changes in the parameter values of the loss function than by changes in the parameter values of the density function. For the latter parameters we can construct confidence intervals (or regions) and use the end points of these intervals as the bounds within which we let the parameter values vary. When using confidence regions the sensitivity of the expected adaptation loss can in the same way be determined for simultaneous changes in the parameter values. It may occur that the width of a confidence interval is that large that at the end points of the interval the difference between the expected adaptation loss of the computed size pattern and the expected adaptation loss of the optimal size pattern becomes unacceptably large. In that case, the confidence interval can be made smaller by increasing the sample size.

Instead of calculating confidence intervals we also may use the Baysian approach. The optimal sample size then can be found by considering the expected loss due to estimation errors and the costs of sampling. Both types of costs are a function of the sample size; the sample size at which the sum of these costs reaches a minimum is the optimal sample size.

Table 11.4. The Percentage Increase of E(L) as a Function of k, B, σ and μ.

Part I: k	E(L) Size Pattern (11.21)	E(L) Optimal Size Pattern	Percentage Difference
.5	.126	.125	.82
.6	.150	.149	.43
.7	.174	.174	.20
.8	.198	.198	.08
.9	.223	.223	.02
1.0	.247	.247	.00
1.1	.271	.271	.01
1.2	.295	.295	.05
1.3	.319	.319	.10

Table 11.4. (Continued)

Part I: k	E(L) Size Pattern (11.21)	E(L) Optimal Size Pattern	Percentage Difference
1.4	.344	.343	.16
1.5	.368	.367	.24
Part II: B			
5	.244	.243	.68
6	.245	.244	.37
7	.245	.245	.18
8	.246	.246	.07
9	.246	.246	.02
10	.247	.247	.00
11	.247	.247	.01
12	.248	.248	.05
13	.248	.248	.11
14	.249	.248	.18
15	.249	.249	.27
Part III: σ			
.5	.186	.125	49.00
.6	.192	.149	28.55
.7	.200	.174	15.08
.8	.211	.198	66.46
.9	.226	.223	1.59
1.0	.247	.247	.00
1.1	.275	.271	1.57
1.2	.313	.295	6.22
1.3	.363	.319	13.69
1.4	.424	.343	23.57
1.5	.496	.367	35.30
Part IV: μ			
5	3.586	.247	1352.84
6	2.589	.247	948.89
7	1.626	.247	558.77
8	.829	.247	235.90
9	.374	.247	51.41
9.5	.277	.247	12.20
10	.247	.247	.00
10.5	.280	.247	13.42
11	.407	.247	65.02
12	1.373	.247	456.23
13	4.165	.247	1587.48
14	7.728	.247	3030.88
15	9.583	.247	3782.70

11.4   THE UNIQUENESS OF A SIZE PATTERN SATISFYING THE RECURSION FORMULAE

In Chap. 4 we have derived a number of formulae the optimal size pattern has to satisfy. These formulae combined with the required positive definiteness of the Hessian—evaluated at the located size pattern—are sufficient for a local minimum.

It is, for a specific case, in general not very difficult to locate a size pattern at which the expected adaptation loss reaches a local minimum. As the word "local," however, already indicates, there might also be other minima. Hence, one of these minima might be smaller than the minimum located, in which case the size pattern found gives only a suboptimal solution.

The existence of two or more local minima depends on the shape of the probability density function of the required sizes and on the form of the loss function used. It can be proved for some combinations of density function and loss function that there exists only one size pattern satisfying the recursion formulae and that the Hessian evaluated at this size pattern is positive definite.

A general proof for the uniqueness of the (located) size pattern, however, cannot be given, for it might occur that the recursion formulae have more than one solution. We have constructed a few examples to illustrate this point. In these examples the linear loss function is used in combination with shift rule O. Furthermore, n is chosen equal to 2, but with the largest size x_2 fixed so that the problem reduces to the choice of an optimal x_1.

Example 1.   Suppose h(x) equals

$$h(x) = \frac{8}{3} x^2 \quad -.5 \leqslant x \leqslant 1,$$

$$= 0 \qquad \text{elsewhere,} \tag{11.22}$$

so that

$$H(x) = \frac{8}{9} x^3 + \frac{1}{9} \quad -.5 \leqslant x \leqslant 1. \tag{11.23}$$

The function h(x) is plotted in Fig. 11.1. Putting k and x_2 both equal to one, x_1 has, according to (4.22), to satisfy the relation:

$$\frac{8}{9} x_1^3 + \frac{1}{9} = (1 - x_1) \frac{8}{3} x_1^2$$

or

$$32 x_1^3 - 24 x_1^2 + 1 = 0. \tag{11.24}$$

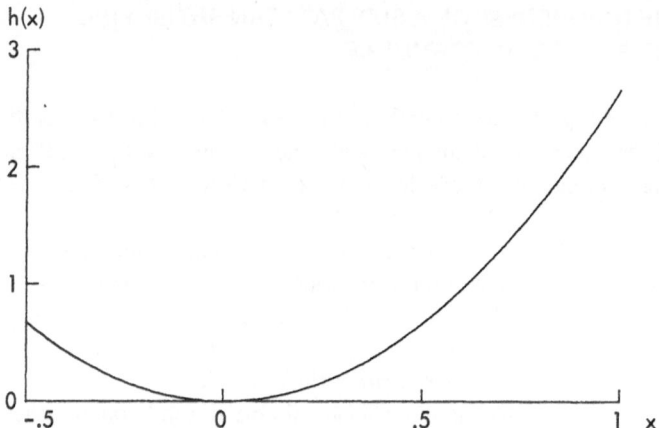

Figure 11.1. The Distribution of the Required Sizes

This equation has the successive roots,

$$x_1 = -.18, \quad x_1 = .25, \quad x_1 = .68.$$

All these roots are in the required interval (-.5, 1); calculating the value of the second derivative of $E(L)$ at these points we find that $x_1 = .25$ is a maximum and that both other roots are minima. In Fig. 11.2 the expected adaptation loss is plotted as a function of x_1. As a result of the chosen density function this function has a rather strange shape. It is namely symmetrical around $x_1 = .25$ so that the expected adaptation losses at $x_1 = -.18$ and $x_1 = .68$ are equal to each other.

If the probability density function has—as in this example—two or maybe even more tops, one must thoroughly take into account the possibility that a solution of the recursion formulae might not be the (strict) local minimum. If no straightforward mathematical solution is possible it is, for this type of density function, recommended to start the iterative method from a large number of starting sets. In this way one can usually easily find all the solutions.

It shall be shown next that if the probability density function is a nondecreasing function of x, there might also exist more than one size pattern satisfying the recursion formulae.

Example 2. Let $h(x)$ for instance be equal to

$$
\begin{aligned}
h(x) &= 1/8 && 0 \leqslant x \leqslant 2, \\
&= 6/8 && 2 < x \leqslant 3, \\
&= 0 && \text{elsewhere.}
\end{aligned}
\qquad (11.25)
$$

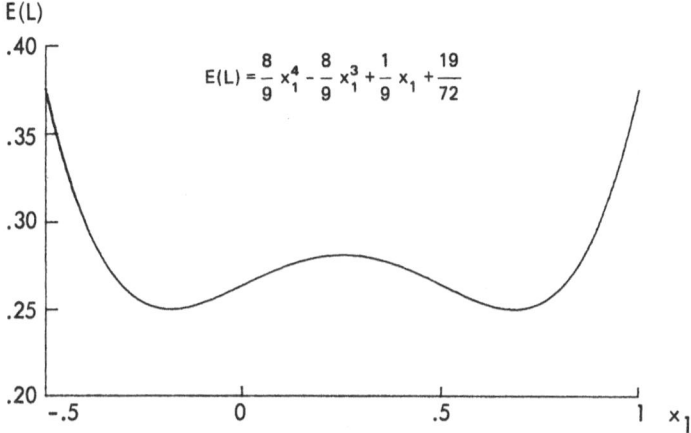

Figure 11.2. The Expected Adaptation Loss as a Function of x_1

If we put x_2 equal to 3 the expected adaptation loss equals,

$$E(L) = \frac{1}{8}\left[x_1^2 - 3x_1 + 7\right] \qquad \text{if } x_1 \leqslant 2,$$

$$= \frac{1}{8}\left[6x_1^2 - 28x_1 + 37\right] \qquad \text{if } x_1 > 2. \qquad (11.26)$$

As easily can be seen, this function possesses two local minima, one at $x_1 = 1.5$ and the other at $x_1 = 2\frac{1}{3}$. The expected adaptation loss at these values equals .59 and .54, respectively.

From the given examples it will be clear now that there might in general be more than one size pattern at which the expected adaptation loss reaches a local minimum. Numerical experimentation however indicates that this complication can be neglected if the probability density function of the required sizes equals one of the four density functions discussed in the Chaps. 5, 6, 7 and 8. For these densities we namely never found more than one solution of the recursion formulae[1]. Also when using other density functions there will usually be one solution, as long as they have one top and a fairly regular shape.

[1] For each of the four density functions it can be proved that size x_i is uniquely determined by the recursion formulae for given size x_{i-1} and x_{i+1}. Though this proof is not sufficient for the existence of a unique solution it at least rules out suboptimal solutions like those discussed in the examples.

12 SIZE PATTERNS WHICH FORM AN ARITHMETICAL OR GEOMETRICAL SERIES

12.1 INTRODUCTION

In practice, size patterns are often chosen according to an arithmetical or geometrical series or an adjusted version of one of these two. In our opinion the strong preference for series of this type is not justified. Apart from the fact that they have no theoretical background, the expected adaptation loss when using these series will usually be considerably larger than that of an optimal size pattern. We will show this by means of some examples in Sec. 12.5. On the other hand, it may occur that a size pattern, chosen according to an arithmetical or geometrical series, is optimal. As we saw in Chap. 5, the use of the homogeneous distribution with a linear loss function led to a size pattern which formed an arithmetical series. An interesting question is whether there exist more of those cases, i.e. is there—when using the linear loss function—a probability density function for which the optimal size pattern is a geometrical series or are there (general) loss functions which lead to arithmetical or geometrical series? In the following sections we will show that such functions indeed exist. The form of these functions, however, will clearly indicate that their field of applicability is very limited.

12.2   A PROBABILITY DENSITY FUNCTION FOR WHICH THE OPTIMAL SIZE PATTERN FORMS A GEOMETRICAL SERIES

In this and the following sections we will assume that shift rule O is being applied and that the largest size, x_n, is fixed before optimization takes place. Thus we do not take into consideration the loss of not being served. When using the linear loss function the recursion formulae are, according to (4.22),

$$\int_{x_{i-1}}^{x_i} h(x)\,dx = (x_{i+1} - x_i)\,h(x_i) \qquad i = 1, \ldots, n-1. \tag{12.1}$$

If the sizes form a geometrical series it follows that

$$x_i / x_{i-1} = r \qquad i = 1, \ldots, n. \tag{12.2}$$

Putting,

$$x_{i-1} = a, \qquad x_i = ar \quad \text{and} \quad x_{i+1} = ar^2 \tag{12.3}$$

we may write (12.1) as

$$\int_a^{ar} h(x)\,dx = (ar^2 - ar)h(ar). \tag{12.4}$$

To find a function h which satisfies (12.4) for every r we differentiate (12.4) with respect to r. This gives,

$$ah(ar) = (2ar - a)h(ar) + a(ar^2 - ar)h'(ar)$$

or

$$h'(ar)/h(ar) = -2/ar \tag{12.5}$$

where $h'(ar)$ denotes the derivative of $h(x)$ with respect to x, evaluated at ar. Substituting $x = ar$, (12.5) reduces to

$$h'(x)/h(x) = -2/x. \tag{12.6}$$

Hence,

$$\frac{\partial}{\partial x}\ln(h(x)) = \frac{\partial}{\partial x}(-2\ln(x)) \tag{12.7}$$

and we have,

$$h(x) = c/x^2 \tag{12.8}$$

where c is a constant, larger than zero. Choosing c equal to x_0, we derive from (12.8) the density function

$$h(x) = x_0/x^2 \quad x \geqslant x_0 \text{ and } x_0 > 0. \tag{12.9}$$

Because,

$$x_n = x_0 r^n \tag{12.10}$$

we have

$$r = (x_n/x_0)^{1/n}. \tag{12.11}$$

Hence, the optimal size pattern forms the geometrical series

$$x_i = x_0 (x_n/x_0)^{i/n} \quad i = 1, \ldots, n. \tag{12.12}$$

The expected adaptation loss is given by

$$E(L) = k \sum_{i=1}^{n} \int_{x_{i-1}}^{x_i} (x_i - x) \frac{x_0}{x^2} dx \tag{12.13}$$

$$= kx_0 \sum_{i=1}^{n} (x_i/x_{i-1} - \ln (x_i/x_{i-1}) - 1) \tag{12.14}$$

which reduces to,

$$E(L) = nkx_0 \left(\left(\frac{x_n}{x_0} \right)^{1/n} - \frac{1}{n} \ln \left(\frac{x_n}{x_0} \right) - 1 \right) \tag{12.15}$$

if the sizes are chosen according to (12.12).

From (12.14) we see that the (total) expected adaptation losses of the subpopulations are equal to each other. Note that this is not the case for the conditional expected adaptation losses of the subpopulations.

The probability density function (12.9) will—because of its rather unusual shape—have little operational significance. Only the shape of the exponential distribution is about the same as the shape of h(x). All the other density functions considered thus far have an entirely different shape; a size pattern chosen according to a geometrical series will therefore in general not be optimal if the linear loss function is used with shift rule O.

12.3 SOME LOSS FUNCTIONS FOR WHICH THE OPTIMAL SIZE PATTERN FORMS A GEOMETRICAL OR AN ARITHMETICAL SERIES

When deriving a loss function for which the optimal size pattern is a geometrical series we may make use of the results of the previous section. Applying namely the transformation,

$$z = 1/x \qquad (12.16)$$

and the substitutions,

$$z_i = 1/x_{n-i} \qquad i = 0, \dots, n \qquad (12.17)$$

to (12.13) we obtain

$$E(L) = \frac{k}{z_n} \sum_{i=1}^{n} \int_{z_{i-1}}^{z_i} \left(\frac{1}{z_{i-1}} - \frac{1}{z} \right) dz$$

$$= \frac{k}{z_n} \sum_{i=1}^{n} \int_{z_{i-1}}^{z_i} \left(\frac{1}{z} - \frac{1}{z_i} \right) dz + \frac{k}{z_n} \left\{ \frac{n(r^2 - 1)}{r} - 2 \ln\left(\frac{z_n}{z_0} \right) \right\} \qquad (12.18)$$

where

$$r = z_i / z_{i-1} = x_{n+1-i} / x_{n-i} \qquad i = 1, \dots, n. \qquad (12.19)$$

The last term of (12.18) does not depend on the sizes to be chosen and can thus be regarded as a constant. The first term of (12.18) equals the expected adaptation loss if the probability density function is the homogeneous distribution on the interval $(z_0, z_0 + z_n)$ and the loss function $k(z^{-1} - z_i^{-1})$ is used in combination with shift rule O. Furthermore, according to (12.19) the optimal size pattern $z_1, \dots, z_n$ forms a geometrical series. Also a more general loss function can be derived, which generates the sizes according to a geometrical series. For this purpose we eliminate the influence of the probability density function from the formula of the expected adaptation loss [see (12.13)] by defining

$$L(x) = k \frac{x_i - x}{x^2 h(x)} \qquad x_{i-1} < x \leqslant x_i, \quad i = 1, \dots, n. \qquad (12.20)$$

Next we will derive a loss function for which the optimal size pattern is an arithmetical series. This might be done by applying the transformation,

$$z = \ln x^2 \qquad (12.21)$$

and the substitutions,

$$z_i = \ln x_i^2 \qquad i = 0, \ldots, n \tag{12.22}$$

to (12.13). We then have

$$E(L) = k \, e^{z_0/2} \sum_{i=1}^{n} \int_{z_{i-1}}^{z_i} (e^{z_i/2} - e^{z/2}) \frac{1}{2} e^{-z/2} \, dz. \tag{12.23}$$

Formula (12.23) arises if the exponential distribution is used with an "exponential loss function." The optimal size pattern $z_1, \ldots, z_n$ forms an arithmetical series because

$$z_i - z_{i-1} = 2 \ln (x_i/x_{i-1}) = 2 \ln r \qquad i = 1, \ldots, n. \tag{12.24}$$

A more general loss function which also generates an arithmetical series is of the form

$$L(x) = k(x_i - x)/h(x) \qquad x_{i-1} < x \leqslant x_i, \ i = 1, \ldots, n. \tag{12.25}$$

The forms of the loss functions specified by (12.20) and (12.25) indicate that these loss functions have no practical importance. Only the exponential loss function given in (12.23) might arise in a practical situation if the price of the product is an exponential function of the property to be standardized. This may for instance occur if the property is the precision of a certain instrument.

12.4   THE LOSS FUNCTION FOR WHICH THE SALES PER SIZE ARE EQUAL TO EACH OTHER

A rule according to which the sizes might be selected is the rule of the equal sales per size. To find a loss function which generates a size pattern satisfying this rule we proceed as follows. Assuming shift rule O to be applied we put the loss function equal to

$$L(x) = k \, (W(x_i) - W(x)) \qquad x_{i-1} < x \leqslant x_i, \ i = 1, \ldots, n. \tag{12.26}$$

The expected adaptation loss equals for fixed x_n

$$E(L) = k \sum_{i=1}^{n} \int_{x_{i-1}}^{x_i} (W(x_i) - W(x))h(x) \, dx. \tag{12.27}$$

Differentiating (12.27) with respect to the sizes we obtain the recursion formulae:

$$W'(x_i)p_i = (W(x_{i+1}) - W(x_i))h(x_i) \quad i = 1, \ldots, n - 1. \quad (12.28)$$

If the function W is chosen equal to H we have

$$H'(x_i)p_i = p_{i+1} h(x_i) \quad i = 1, \ldots, n - 1.$$

Hence,

$$p_i = p_{i+1} \quad i = 1, \ldots, n - 1. \quad (12.29)$$

From (12.29) we see that, if the sizes are chosen optimally, the probabilities of the service areas are equal to each other. The probability of each service area then equals the coverage of the size pattern divided by n. Furthermore, if the size pattern satisfies (12.29), the expected adaptation loss is given by

$$E(L) = \frac{1}{2}k\, p^2/n \quad (12.30)$$

where p is the coverage of the size pattern. Except if the homogeneous distribution is used, the loss function (12.26) with the distribution function H substituted for W, is not very likely to arise in a practical situation. Therefore, also a size pattern chosen with the rule of equal sales per size will usually not be economically optimal, at least not if shift rule O is being applied.

12.5 COMPARISON OF AN OPTIMAL SIZE PATTERN WITH SIZE PATTERNS CHOSEN ACCORDING TO AN ARITHMETICAL OR GEOMETRICAL SERIES

To illustrate the differences between an optimal size pattern and size patterns chosen according to an arithmetical or geometrical series, we have inserted into Table 12.1 some size patterns together with their expected adaptation losses for n ranging from 1 to 5. For the distribution of the required sizes we took the normal distribution with $\mu = 10$ and $\sigma = 2$. Furthermore, the linear loss function with $k = 1$ was used in combination with shift rule O. To facilitate comparisons between the various size patterns we have fixed x_0 and x_n at the respective values of $(\mu - 3\sigma)$ and $(\mu + 3\sigma)$ and neglected the demand outside this range. The geometrical series appears to have the largest expected adaptation loss for all values of n larger than unity. This is caused by the fact that when using this series (with $n > 1$) relatively many sizes are located below the mean of the distribution as opposed to the optimal series where most sizes are located

Table 12.1. The Optimal Size Patterns and the Size Patterns Chosen According to the Arithmetical and Geometrical Series with the Associated Expected Adaptation Losses for n Ranging from 1 to 5 (Linear Loss Function)

	Size/E(L)	Optimal Series	Arithmetical Series	Geometrical Series
n = 1	1	16	16	16
	E(L)	5.98	5.98	5.98
n = 2	1	11.33	10	8
	2	16	16	16
	E(L)	2.50	2.99	4.73
n = 3	1	9.88	8	6.35
	2	12.26	12	10.08
	3	16	16	16
	E(L)	1.60	1.99	2.82
n = 4	1	9.07	7	5.66
	2	10.86	10	8
	3	12.76	13	11.31
	4	16	16	16
	E(L)	1.18	1.50	1.95
n = 5	1	8.54	6.4	5.28
	2	10.05	8.8	6.96
	3	11.44	11.2	9.19
	4	13.10	13.6	12.13
	5	16	16	16
	E(L)	0.94	1.20	1.52

above the mean. The expected adaptation loss of the arithmetical series, which also has most sizes above the mean if $n > 2$, is between the expected adaptation losses of the other series for all values of n.

Table 12.2 shows the results if instead of the linear loss function the quadratic loss function is used. As readily can be verified, the relative differences between the expected adaptation losses of the optimal series and the other two series—which were already considerable for the linear loss function—have become still even larger.

Table 12.2. The Optimal Size Patterns and the Size Patterns Chosen According to the Arithmetical and Geometrical Series with the Associated Expected Adaptation Losses for n Ranging from 1 to 5 (Quadratic Loss Function)

	Size/E(L)	Optimal Series	Arithmetical Series	Geometrical Series
n = 1	1	16	16	16
	E(L)	151.70	151.70	151.70
n = 2	1	11.56	10	8
	2	16	16	16
	E(L)	56.13	73.91	121.49
n = 3	1	10.13	8	6.35
	2	12.49	12	10.08
	3	16	16	16
	E(L)	34.48	45.03	70.26
n = 4	1	9.34	7	5.66
	2	11.13	10	8
	3	12.99	13	11.31
	4	16	16	16
	E(L)	24.94	32.88	46.09
n = 5	1	8.82	6.4	5.28
	2	10.34	8.8	6.96
	3	11.71	11.2	9.19
	4	13.32	13.6	12.13
	5	16	16	16
	E(L)	19.55	25.83	34.77

Finally, we will briefly pay some attention to an interesting practical example, which, once more, clearly illustrates the serious disadvantages the use of a geometrical series, or more specifically a preferred number series, may have. The example—concerning the standardization of the nominal diameters (in centimetres) of saucepans—is originally given by Sanders (1972) with the purpose to point out the advantages of the preferred number series.

In a market survey on saucepans, referred to by Sanders, the following frequency distribution of the required diameters was obtained (see Fig. 12.1). The continuous curve in Fig. 12.1 represents a log normal distribution. Its parameter

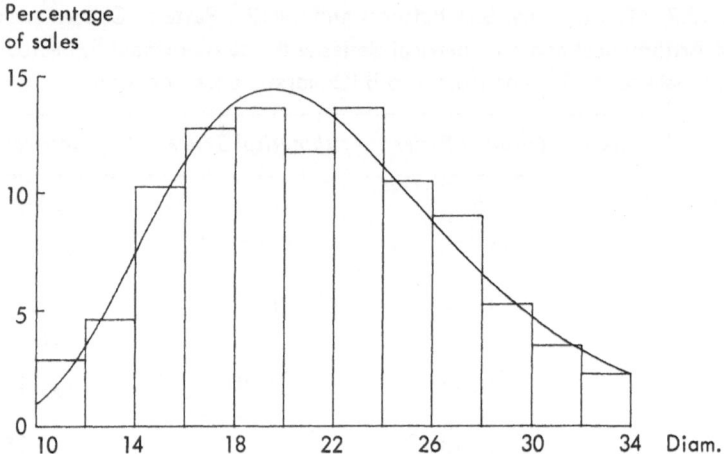

Figure 12.1. The Distribution of the Required Diameters

values (i.e., $\hat{\mu}$ = 3.053 and $\hat{\sigma}$ = .2878) were estimated by employing the maximum likelihood technique discussed in Chap. 10. The computed chi square value indicates that the distribution gives a good fit to the observed frequencies.

The parameter estimation above has been carried out under the assumption that housewives apply shift rule O when buying saucepans. This assumption is motivated by the fact that —when confronted with the choice between a smaller and a larger diameter if the required diameter is not available— a housewife usually will buy the larger diameter in order to have some overcapacity. Since the thickness of the bottom and the wall of saucepans in general does not vary with the height and the diameter we may express the adaptation loss in cm^2 of the raw material used. The surface of a pan with diameter x and height y equals $\frac{1}{4}\pi x^2 + \pi xy$ cm^2. If we assume the height to be proportional to the diameter, say y = cx, the adaptation loss can be written as

$$L(x) = (\tfrac{1}{4}\,\pi + c\pi)(x_i^2 - x^2) \qquad x_{i-1} < x \leqslant x_i, \ i = 1, \ldots, n. \quad (12.31)$$

For the further computations we have chosen c equal to .6. Note that this choice does not influence the position of the optimal size pattern nor the relative difference between the expected adaptation losses of two different size patterns.

The size pattern for the diameters of saucepans, derived by Sanders from the R10 series, consists of 5 sizes. These are inserted into Table 12.3. The largest size of the size pattern is 31.5 cm which leads to a coverage of .916. Neglecting the demand above 31.5 cm —which implies that the loss of not being served is

Table 12.3. The Size Pattern Chosen According to the Preferred Number Series and the Optimal Size Pattern

Size No.	Preferred Numbers	Prob. Serv. Areas	Optimal Sizes	Prob. Serv. Areas
1	12.5	.036	17.25	.260
2	16	.144	20.73	.253
3	20	.280	23.95	.213
4	25	.324	27.40	.163
5	31.5	.216	31.50	.111
E(L)	308.28		225.95	

not taken into consideration—we may compute the (conditional) expected adaptation loss of a consumer, given that he (or she) is served, by means of

$$E(L) = \frac{2.670 \sum_{i=1}^{5} \int_{x_{i-1}}^{x_i} (x_i^2 - x^2)h(x)dx}{.916} = 308.28 \text{ cm}^2. \qquad (12.32)$$

The average required surface of a pan is,

$$\frac{2.670 \int_0^{x_n} x^2 h(x)dx}{.916} = 1217.80 \text{ cm}^2 \qquad (12.33)$$

so that the expected adaptation loss—expressed as a percentage of the average weight of a pan—equals 25.31%.

Next, the optimal size pattern consisting of 5 sizes—with the largest size equaling 31.5 cm—has been calculated. Table 12.3 shows the optimal sizes and the probabilities of the service areas. The expected adaptation loss of this size pattern equals 225.95 cm^2. From the table we see that the expected adaptation loss of the size pattern based on the preferred number series is 82.33 cm^2 or 36.44% larger than this amount. The expected adaptation loss of the optimal size pattern—expressed as a percentage of (12.33)—equals 18.55%. The average weight of saucepans and the energy required to heat the pans can therefore be reduced by 6.7% by making use of the optimal size pattern instead of the size pattern proposed by Sanders.

Figure 12.2 shows the percentage increase of the expected adaptation loss for some other values of n, if instead of the optimal size pattern a geometrical or arithmetical series is chosen. The smallest and the largest size of the latter two series were—as in the example above—put equal to respectively 12.5 cm and 31.5 cm. As can be seen, even for larger values of n, the savings due to the choice of an optimal size pattern are considerable. This is especially the case if the geometrical series is used; the expected adaptation losses associated with the arithmetical series are—as in the previous examples—much closer to those of the optimal size patterns.

Examining the results of the examples above, we conclude that it is not to be recommended to use mechanically a geometrical or arithmetical series to select the size pattern. Although in some instances these series might be optimal, this is usually not so and in that case some further research directed to an optimally located size pattern might reduce the expected adaptation loss considerably.

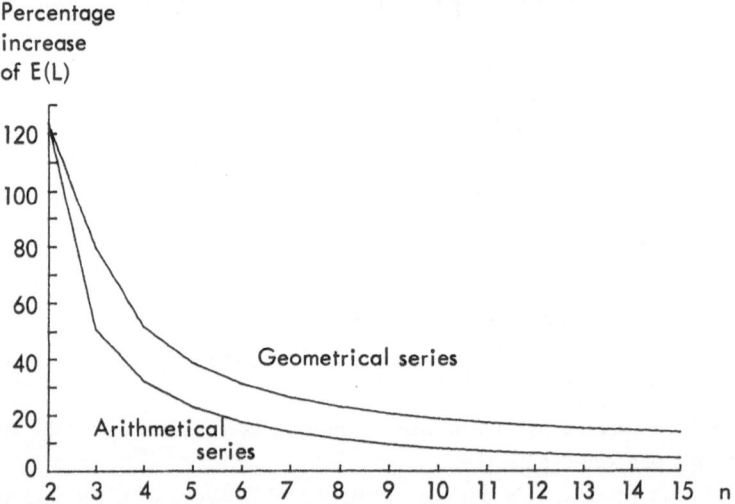

Figure 12.2. The Percentage Increase of E(L) If the Geometrical or Arithmetical Series Instead of the Optimal Series Is Used

13 THE STANDARDIZATION OF TWO PROPERTIES OF A PRODUCT

13.1 INTRODUCTION

Thus far we have only taken into consideration products of which one property was to be standardized. However, it will often occur that standardization of two or more properties is necessary. We may for instance think of the inner and outer diameters of pipes or the power and the number of revolutions of electro motors. In this and the following chapters attention will be paid to standardization problems where two properties are to be standardized. If we write the required size as the vector (x,y), where x and y are the required values of the respective product properties and the ith size as (x_i, y_i), then the problem is to find n sizes (x_i, y_i) $(i = 1, \ldots, n)$ such that they minimize the expected adaptation loss. As the choice of n sizes is equivalent to the choice of n points in the two dimensional space, it will be clear that the service areas of the sizes are now—instead of intervals—regions. The surface as well as the shape of the regions will depend on the shift rule, the loss function and the size pattern chosen. Note that in the one dimensional case only the width of the interval and not the shape could change. It became apparent that the different shapes the service areas may have, cause major problems when standardizing two properties.

131

To avoid these problems we will begin by successively introducing three restrictions on the relative position of the sizes. The restrictions are respectively named the grid-, the line- and the generalized grid restriction; all three restrictions have the effect that—though the probabilities of the service areas may differ—the shapes of the service areas become equal to each other. In a later chapter we will discuss unrestricted selection of the sizes.

Two important elements in any standardization problem are the loss function and the probability density function of the required sizes. We will consider two loss functions in Sec. 13.2; the first loss function is a generalization of the one used when standardizing one property whereas the second is specifically designed for the two (or more) dimensional case. As opposed to the one dimensional problem we will now use only one shift rule, namely shift rule O. This implies that a consumer with demand (x,y) has to purchase a size (x_i, y_i) such that $x_i \geqslant x$ and $y_i \geqslant y$. In case no size (x_i, y_i) $(i = 1, \ldots, n)$ satisfies this requirement, the consumer is assumed to buy no standardized size at all; he then suffers a loss of not being served which we put—as in the one dimensional case—equal to a constant, denoted by the letter B.

Section 13.3 contains the probability density functions that are used in the subsequent chapters. These are the bivariate homogeneous distribution, the bivariate normal distribution and the bivariate log normal distribution. The last two density functions are—contrary to the first—able to cope with situations where the demands for the product properties are correlated and are therefore the most important.

13.2 THE LOSS FUNCTIONS

13.2.1 Loss Function I

Loss function I is a direct generalization of (2.8). It has the form

$$
\begin{aligned}
L(x,y) = {}& \underset{\{i \,|\, x_i \,\geqslant\, x \text{ and } y_i \,\geqslant\, y\}}{\mathrm{Min}} k_1 (x_i^{\alpha_1} - x^{\alpha_1})^{\beta_1} \\
& + k_2 (y_i^{\alpha_2} - y^{\alpha_2})^{\beta_2} \quad \alpha_1, \alpha_2, \beta_1, \beta_2 > 0 \\
= {}& B \quad \text{if there is no size } (x_i, y_i) \text{ such that } x_i \geqslant x \text{ and} \\
& y_i \geqslant y \quad (i = 1, \ldots, n).
\end{aligned}
\tag{13.1}
$$

The variables x_i and y_i in (13.1) are the coordinates of the ith size and n is the number of sizes to be chosen. As can be seen the loss function consists of two parts. The first part gives the adaptation loss caused by the adaptation of the first property while the second part does the same for the second property. Consequently, the loss function can be used only in situations where these losses are independent. Furthermore, when standardizing two properties, there may in principle be more sizes satisfying the requirements implied by the use of shift rule O. It is assumed that the consumer chooses of these sizes the size which leads to the smallest adaptation loss. This adaptation loss is determined by means of the minimizing operation in (13.1). The loss function (13.1) will for instance arise if steel beams with a predescribed resistance moment in the length direction, x, and in the width direction, y, are required for a certain construction. If a standardized size (x_i, y_i) is used for this construction than the adaptation loss in kg per running meter equals [Sittig, (1973)]

$$L(x,y) = 0.38 \, (x_i^{2/3} - x^{2/3}) + 0.74 \, (y_i^{2/3} - y^{2/3}). \qquad (13.2)$$

In the next chapters, most attention will be paid to the loss function which arises if we put the parameters α_1, α_2, β_1 and β_2 equal to unity. The adaptation loss is then a linear function of the required values and the purchased values of the product properties; we will call this loss function therefore the linear loss function.

13.2.2 Loss Function II

The second loss function we will use is defined by

$$L(x,y) = \min_{\{i \, | \, x_i \, \geqslant \, x \text{ and } y_i \, \geqslant \, y\}} k(x_i^{\alpha_1} y_i^{\alpha_2} - x^{\alpha_1} y^{\alpha_2}) \qquad \alpha_1, \alpha_2 > 0,$$

$$= B \quad \text{if there is no size } (x_i, y_i) \text{ such that } x_i \geqslant x \text{ and}$$
$$y_i \geqslant y \quad (i = 1, \ldots, n). \qquad (13.3)$$

The loss function (13.3) might be of use in a wide variety of standardization problems. We mention for example the standardization of window frames where a loss function of this type was used [Wouters, (1969)]. The loss function which arises if we put $\alpha_1 = \alpha_2 = 1$ shall be studied more intensively. As this loss function has, loosely speaking, a quadratic character we will call it the quadratic loss function. It for instance can be used to describe the adaptation loss when standardizing the thickness and width of flat steel sections.

13.3 THE PROBABILITY DENSITY FUNCTIONS

13.3.1 The Bivariate Homogeneous Distribution

The density function of the bivariate homogeneous distribution is,

$$s(x,y) = \frac{1}{(b-a)(d-c)} \quad \text{if } a \leqslant x \leqslant b, \, c \leqslant y \leqslant d,$$

$$= 0 \qquad \qquad \text{elsewhere}, \qquad (13.4)$$

where a and c are supposed to be nonnegative. The symbols x_0 and x_{n+1} are defined by (4.1) and (4.2); the function h(x) in these expressions denotes the marginal density function of x. Defining y_0 and y_{n+1} similarly, we have

$$x_0 = a, \quad x_{n+1} = b, \quad y_0 = c \quad \text{and} \quad y_{n+1} = d.$$

The mean and variance of x are given by (5.2). Substituting in (5.2) c for a and d for b we obtain the mean and variance of y.

13.3.2   The Bivariate Normal Distribution

The bivariate normal distribution is denoted by the letter w. Its probability density function is

$$w(x,y) = \frac{1}{2\pi\sqrt{1-\rho^2}\,\sigma_x\sigma_y}\exp\left[-\left(\left(\frac{x-\mu_x}{\sigma_x}\right)^2 - 2\rho\left(\frac{x-\mu_x}{\sigma_x}\right)\left(\frac{y-\mu_y}{\sigma_y}\right)\right.\right.$$
$$\left.\left.+\left(\frac{y-\mu_y}{\sigma_y}\right)^2\right)\bigg/(2(1-\rho^2))\right] \quad -\infty < x,y < \infty. \tag{13.5}$$

The parameters μ_x, μ_y, σ_x and σ_y are respectively the means and the standard deviations of x and y. The parameter ρ denotes the correlation coefficient; it will be assumed in the next chapters that ρ has a nonnegative value. Furthermore, referring to the remark concerning the normal distribution in Sec. 4.2, $x_0 = y_0 = 0$ and $x_{n+1} = y_{n+1} = \infty$. The transformed variables,

$$v = (x - \mu_x)/\sigma_x \quad \text{and} \quad u = (y - \mu_y)/\sigma_y \tag{13.6}$$

have the bivariate standard normal distribution with the density function

$$g(v,u) = \frac{1}{2\pi\sqrt{1-\rho^2}}e^{-\frac{1}{2}\{v^2 - 2\rho vu + u^2\}/(1-\rho^2)} \quad -\infty < v,u < \infty. \tag{13.7}$$

The distribution function of the bivariate standard normal distribution is

$$G(a,b) = \int_{-\infty}^{a}\int_{-\infty}^{b} g(v,u)\,du\,dv. \tag{13.8}$$

When using the bivariate normal- or log normal distribution we will reduce all the double integrals in the recursion formulae to the form (13.8). This has the advantage that only one numerical routine is required to calculate them. We used in this routine the series expansion (26.3.29) given by Abramowitz (1972) on page 940.

13.3.3 The Bivariate Log Normal Distribution

As opposed to the bivariate normal distribution the bivariate log normal distribution is only defined in the positive quadrant. The probability density function equals

$$
z(x,y) = \frac{1}{xy2\pi\sqrt{1 - \rho^2}\,\sigma_1\sigma_2} \exp\left[-\left(\left(\frac{\ln x - \mu_1}{\sigma_1}\right)^2\right.\right.
$$
$$
\left.\left. - 2\rho\left(\frac{\ln x - \mu_1}{\sigma_1}\right)\left(\frac{\ln y - \mu_2}{\sigma_2}\right)\right.\right.
$$
$$
\left.\left. + \left(\frac{\ln y - \mu_2}{\sigma_2}\right)^2\right)/(2(1 - \rho^2))\right] \quad x,y > 0,
$$
$$
= 0 \qquad\qquad\qquad \text{elsewhere.} \quad (13.9)
$$

The means and variances of x and y are

$$
E(x) = e^{\mu_1 + \sigma_1^2/2}, \qquad\qquad E(y) = e^{\mu_2 + \sigma_2^2/2},
$$
$$
V(x) = e^{2\mu_1 + \sigma_1^2}(e^{\sigma_1^2} - 1), \quad V(y) = e^{2\mu_2 + \sigma_2^2}(e^{\sigma_2^2} - 1). \quad (13.10)
$$

Furthermore, $x_0 = y_0 = 0$ and $x_{n+1} = y_{n+1} = \infty$. To evaluate integrals of the form,

$$
Z(a,b) = \int_0^a \int_0^b z(x,y)\,dy\,dx \quad (13.11)
$$

we apply the transformations,

$$
v = \frac{\ln x - \mu_1}{\sigma_1} \quad \text{and} \quad u = \frac{\ln y - \mu_2}{\sigma_2} \quad (13.12)
$$

to (13.11). The integral then reduces to the form (13.8) since the variables v and u have bivariate standard normal distribution with correlation coefficient ρ.

14 THE GRID RESTRICTION

14.1 INTRODUCTION

The grid restriction is the first of the three restrictions we will introduce. As its name already suggests the restriction requires that the size pattern forms a grid of the type indicated by Fig. 14.1. In the figure the sizes are located at the intersection points of the horizontal and vertical lines. The grid thus contains 20 sizes.

Each size has an x-coordinate and a y-coordinate; the subscripts of the coordinates of a size are related to the relative position of the size in the grid. Size (x_1, y_1) is for instance always the size in the left lower corner of the grid. Note that in the previous chapter the subscripts were employed in another fashion. When using the grid restriction, we will call the separate coordinates $x_1, x_2, \ldots, y_1, y_2, \ldots$, standardization points. From the figure we see that 5 standardization points were chosen with respect to the first property (x) and 4 standardization points with respect to the second property (y). In this chapter we will assume that n standardization points have to be chosen with respect to the first property and m standardization points with respect to the second property. This leads to a size pattern which contains nm sizes.

136

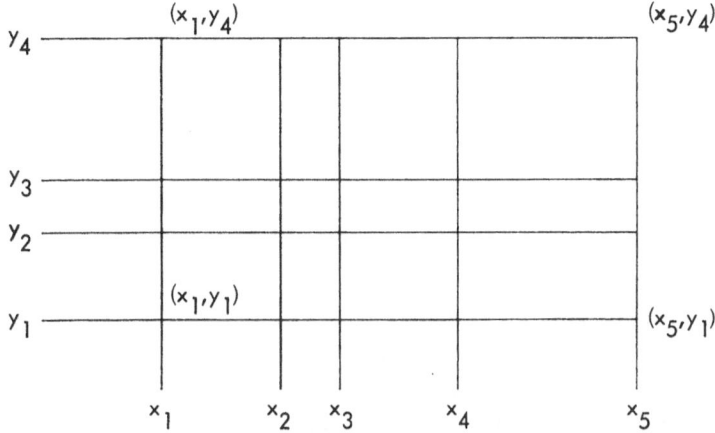

Figure 14.1. An Example of a Grid with n = 5 and m = 4

The main advantage of the use of the grid restriction is that the determination of the service areas becomes rather easy.

From the fact that shift rule O is applied and from the form of the loss functions (13.1) and (13.3) it namely follows that a consumer whose required size is a point in one of the rectangles—induced by the required structure of the size pattern—will purchase the standardized size which is located at the right upper corner of the rectangle. The service area of the size (x_i, y_j) thus equals the rectangle of which the point (x_i, y_j) is the right upper corner.

In the next two sections we will derive the recursion formulae for loss function I and loss function II. The subsequent sections contain the recursion formulae for two specific density functions. These are illustrated with some numerical examples.

14.2 THE RECURSION FORMULAE FOR LOSS FUNCTION I

When using the grid restriction with loss function I we can write the expected adaptation loss as

$$E(L) = \sum_{i=1}^{n} \sum_{j=1}^{m} \int_{x_{i-1}}^{x_i} \int_{y_{j-1}}^{y_j} \{k_1(x_i^{\alpha_1} - x^{\alpha_1})^{\beta_1}$$

$$+ k_2(y_j^{\alpha_2} - y^{\alpha_2})^{\beta_2}\} \, h(x,y) \, dy \, dx$$

$$+ B \left(1 - \int_{x_0}^{x_n} \int_{y_0}^{y_m} h(x,y) \, dy \, dx \right). \qquad (14.1)$$

To find the optimal size pattern we differentiate (14.1) with respect to the $n + m$ standardization points, $x_1, x_2, \ldots, x_n, y_1, \ldots, y_m$. The derivatives with respect to $x_1, \ldots, x_n$ are

$$\frac{\partial E(L)}{\partial x_i} = k_1 \beta_1 \alpha_1 x_i^{\alpha_1 - 1} \int_{x_{i-1}}^{x_i} \int_{y_0}^{y_m} (x_i^{\alpha_1} - x^{\alpha_1})^{\beta_1 - 1} h(x,y) \, dy \, dx$$

$$- k_1 (x_{i+1}^{\alpha_1} - x_i^{\alpha_1})^{\beta_1} \int_{y_0}^{y_m} h(x_i, y) \, dy \qquad i = 1, \ldots, n-1, \quad (14.2)$$

$$\frac{\partial E(L)}{\partial x_n} = k_1 \beta_1 \alpha_1 x_n^{\alpha_1 - 1} \int_{x_{n-1}}^{x_n} \int_{y_0}^{y_m} (x_n^{\alpha_1} - x^{\alpha_1})^{\beta_1 - 1} h(x,y) \, dy \, dx$$

$$+ k_2 \sum_{j=1}^{m} \int_{y_{j-1}}^{y_j} (y_j^{\alpha_2} - y^{\alpha_2})^{\beta_2} h(x_n, y) \, dy - B \int_{y_0}^{y_m} h(x_n, y) \, dy. \quad (14.3)$$

It is not necessary to derive the derivatives with respect to $y_1, \ldots, y_m$ separately since these have the same form as the derivatives with respect to $x_1, \ldots, x_n$ so that they can easily be obtained from (14.2) and (14.3). Putting (14.2), (14.3) and the derivatives with respect to $y_1, \ldots, y_m$ equal to zero we have the recursion formulae. From (14.2) and (14.3) we see that it will in general be necessary to use a numerical integration routine, like Simsons rule, to solve this set of equations. The double integral appearing in the first term of (14.2) for instance cannot be computed directly in most cases. To avoid unnecessary and lengthy calculations we have circumvented the use of numerical integration procedures by choosing in the examples the values of $\alpha_1, \alpha_2, \beta_1$ and β_2 such that the integrals can be calculated directly or can be reduced to the form (13.8). There are various methods which can be used to solve the recursion formulae. We will briefly describe a simple method, based on the method of interval bisection, which proved to work satisfactorily. When using this method, we first choose a starting set, $y_1^1, y_2^1, \ldots, y_m^1$. Given this set, the set $x_1^1, x_2^1, \ldots, x_n^1$, satisfying (14.2) and (14.3), is calculated with the method of interval bisection which is described in Sec. 4.6.2. Note that application of interval bisection causes no problems since, according to (14.2), x_{i+1} can be expressed as an explicit function of x_i, x_{i-1} and y_m. The second step consists of the calculation of $y_1^2, \ldots, y_m^2$, given the calculated set $x_1^1, \ldots, x_n^1$. This also can be done with the method of interval bisection. Next the process is repeated until both sets simultaneously satisfy the nth and (n + m)th recursion formula with a certain precision, say q. The procedure might be accelerated by selecting q rather large at the beginning of the process and letting it decrease after each step, until it reaches a predetermined value. It is recommended to compute the expected adaptation losses of the calculated size patterns after each step and print these together with the size patterns themselves. In this way one can check whether the ultimate value of q is sufficiently low. If this value is namely chosen too large there will be still significant changes of the E(L) or of the sizes in the final iterations in which case some more iterations have to be performed.

14.3 THE RECURSION FORMULAE FOR LOSS FUNCTION II

When using loss function II, the expected adaptation loss can be written as

$$E(L) = k \sum_{i=1}^{n} \sum_{j=1}^{m} \int_{x_{i-1}}^{x_i} \int_{y_{j-1}}^{y_j} (x_i^{\alpha_1} y_j^{\alpha_2} - x^{\alpha_1} y^{\alpha_2}) \, h(x,y) \, dy \, dx$$

$$+ B \left(1 - \int_{x_0}^{x_n} \int_{y_0}^{y_m} h(x,y) \, dy \, dx \right). \tag{14.4}$$

The derivatives with respect to $x_1, \ldots, x_n$ are

$$\frac{\partial E(L)}{\partial x_i} = k \alpha_1 x_i^{\alpha_1 - 1} \sum_{j=1}^{m} y_j^{\alpha_2} \int_{x_{i-1}}^{x_i} \int_{y_{j-1}}^{y_j} h(x,y) \, dy \, dx$$

$$- k(x_{i+1}^{\alpha_1} - x_i^{\alpha_1}) \sum_{j=1}^{m} y_j^{\alpha_2} \int_{y_{j-1}}^{y_j} h(x_i,y) \, dy \qquad i = 1, \ldots, n - 1, \tag{14.5}$$

$$\frac{\partial E(L)}{\partial x_n} = k \alpha_1 x_n^{\alpha_1 - 1} \sum_{j=1}^{m} y_j^{\alpha_2} \int_{x_{n-1}}^{x_n} \int_{y_{j-1}}^{y_j} h(x,y) \, dy \, dx$$

$$+ k x_n^{\alpha_1} \sum_{j=1}^{m} \int_{y_{j-1}}^{y_j} (y_j^{\alpha_2} - y^{\alpha_2}) \, h(x_n,y) \, dy - B \int_{y_0}^{y_m} h(x_n,y) \, dy. \tag{14.6}$$

The derivatives with respect to $y_1, \ldots, y_m$ have a similar form.

Clearly, the recursion formulae for this loss function are easier to solve than those which arise when using loss function I. As namely can be seen, none of the double integrals in (14.5) and (14.6) requires the use of a numerical integration method if the function h equals one of the three density functions mentioned in Chap. 13. Only the integral,

$$\int_{y_{j-1}}^{y_j} y^{\alpha_2} \, h(x_n,y) \, dy$$

appearing in the second term of (14.6) cannot be computed directly for arbitrary values of α_2 if the function h equals the bivariate normal distribution.

To solve the recursion formulae we may use the same technique as in Sec. 14.2. For the specific cases that the log normal distribution is used or the normal distribution with $\alpha_1 = \alpha_2 = 1$ all the integrals appearing in the formula of the expected adaptation loss and in the recursion formulae can be reduced to the form (13.8) or to integrals of the one dimensional standard normal distribution. This reduces the programming effort considerably since most statistical software packages contain subroutines which are able to compute these integrals. The results of the derivations are given in Appendix B.

14.4   THE RECURSION FORMULAE FOR SOME SPECIFIC DISTRIBUTIONS AND LOSS FUNCTIONS

14.4.1   The Bivariate Homogeneous Distribution and Loss Function I

When using the homogeneous distribution and loss function I with $\alpha_1 = \alpha_2 = 1$ there exists an explicit solution of the recursion formulae. This solution can be obtained in the following way. Substituting (13.4) for h in (14.2) and (14.3) and evaluating the integrals we have

$$\frac{\partial E(L)}{\partial x_i} = \{ k_1(y_m - y_0)(x_i - x_{i-1})^{\beta_1}$$
$$- k_1(x_{i+1} - x_i)^{\beta_1}(y_m - y_0)\}/e \quad i = 1, \ldots, n-1, \quad (14.7)$$

$$\frac{\partial E(L)}{\partial x_n} = \left\{ k_1(y_m - y_0)(x_n - x_{n-1})^{\beta_1} \right.$$
$$\left. + \frac{k_2}{\beta_2 + 1} \sum_{j=1}^{m} \{(y_j - y_{j-1})^{\beta_2 + 1}\} - B(y_m - y_0) \right\} /e \quad (14.8)$$

where e equals $(b - a)(d - c) = (x_{n+1} - x_0)(y_{m+1} - y_0)$. Putting (14.7) equal to zero we obtain

$$x_{i+1} - x_i = x_i - x_{i-1} = r_1 \quad i = 1, \ldots, n-1. \quad (14.9)$$

The similar result must be valid for $y_1, \ldots, y_m$. Hence

$$y_{j+1} - y_j = y_j - y_{j-1} = r_2 \quad j = 1, \ldots, m-1. \quad (14.10)$$

Substituting (14.9) and (14.10) into (14.8) and putting (14.8) equal to zero gives,

$$k_1 mr_2 r_1^{\beta_1} + (k_2/(\beta_2 + 1))mr_2^{\beta_2+1} - Bmr_2 = 0$$

or

$$k_1 r_1^{\beta_1} + (k_2/(\beta_2 + 1))r_2^{\beta_2} - B = 0. \tag{14.11}$$

From the derivative of the expected adaptation loss with respect to y_m we obtain in the same way

$$k_2 r_2^{\beta_2} + (k_1/(\beta_1 + 1))r_1^{\beta_1} - B = 0. \tag{14.12}$$

Solving (14.11) and (14.12) for r_1 and r_2 we have

$$r_1 = \left[\frac{B\beta_2(\beta_1 + 1)}{k_1(\beta_1\beta_2 + \beta_1 + \beta_2)} \right]^{\frac{1}{\beta_1}}, \quad r_2 = \left[\frac{B\beta_1(\beta_2 + 1)}{k_2(\beta_1\beta_2 + \beta_1 + \beta_2)} \right]^{\frac{1}{\beta_2}}. \tag{14.13}$$

Note that (14.13) is only a feasible solution if

$$r_1 \leqslant (x_{n+1} - x_0)/n \quad \text{and} \quad r_2 \leqslant (y_{m+1} - y_0)/m. \tag{14.14}$$

If for instance r_1, calculated according to (14.13), violates (14.14) we put r_1 equal to its maximum value and calculate r_2 by means of (14.11). The expected adaptation loss can be calculated by substituting (13.4), (14.9) and (14.10) into (14.1). This leads to

$$
\begin{aligned}
E(L) = & \left\{ \sum_{i=1}^{n} \sum_{j=1}^{m} \frac{k_1}{\beta_1 + 1} (y_j - y_{j-1})(x_i - x_{i-1})^{\beta_1 + 1} \right. \\
& \left. + \frac{k_2}{\beta_2 + 1} (x_i - x_{i-1})(y_j - y_{j-1})^{\beta_2 + 1} \right\} / e \\
& + B \{1 - (y_m - y_0)(x_n - x_0)/e\} \\
= & \frac{nmr_1 r_2}{e} \left\{ \frac{k_1}{\beta_1 + 1} r_1^{\beta_1} + \frac{k_2}{\beta_2 + 1} r_2^{\beta_2} \right\} + B(1 - nmr_1 r_2/e). \tag{14.15}
\end{aligned}
$$

Finally, substitution of (14.13) into (14.15) gives the expected adaptation loss of an optimally chosen size pattern

$$
\begin{aligned}
E(L) = & \frac{nmr_1 r_2}{e} \left\{ \frac{B\beta_2 + B\beta_1}{\beta_1\beta_2 + \beta_1 + \beta_2} \right\} + B(1 - nmr_1 r_2/e) \\
= & B \left\{ 1 - \frac{nmr_1 r_2 \beta_1 \beta_2}{(\beta_1\beta_2 + \beta_1 + \beta_2)e} \right\} \tag{14.16}
\end{aligned}
$$

if the conditions (14.14) are satisfied. If, on the other hand, the optimal values of r_1 and r_2 equal their maximum permittable values, the expected adaptation loss equals, according to (14.15)

$$E(L) = \frac{k_1}{\beta_1 + 1} \left(\frac{x_{n+1} - x_0}{n} \right)^{\beta_1} + \frac{k_2}{\beta_2 + 1} \left(\frac{y_{m+1} - y_0}{m} \right)^{\beta_2}. \tag{14.17}$$

When putting $k_1 = k_2 = \beta_1 = \beta_2 = 1, x_{n+1} - x_0 = y_{m+1} - y_0 = b - a$ and $n = m$, (14.17) reduces to

$$E(L) = k(b - a)/n. \tag{14.18}$$

A similar formula was found in Sec. 5.4 [see (5.13)] where the homogeneous distribution and the linear loss function were used in combination with shift rule O. Note, however, that now — instead of $n - n^2$ sizes are chosen. This implies that, when standardizing two product properties instead of one, the optimal number of sizes might — depending on the behavior of the production and distribution cost — increase considerably.

From (14.17) the optimal values of n and m can be derived, given the number of sizes, say q, that has to be chosen. Substituting $m = q/n$ in (14.17) and putting the derivative with respect to n equal to zero, we obtain

$$n^{opt} = \left[\frac{k_1 \beta_1 (x_{n+1} - x_0)^{\beta_1} (\beta_2 + 1)}{k_2 \beta_2 (y_{m+1} - y_0)^{\beta_2} (\beta_1 + 1)} q^{\beta_2} \right]^{1/(\beta_1 + \beta_2)}. \tag{14.19}$$

If the loss function is linear, this expression reduces to

$$n^{opt} = \sqrt{\frac{k_1 (x_{n+1} - x_0)}{k_2 (y_{m+1} - y_0)} q}. \tag{14.20}$$

Clearly, n^{opt} increases if k_1, the range of x (i.e. $x_{n+1} - x_0 = b - a$) or the number of sizes becomes larger. Since the standard deviations of x and y equal respectively,

$$s_x = (x_{n+1} - x_0)/\sqrt{12} \quad \text{and} \quad s_y = (y_{m+1} - y_0)/\sqrt{12} \tag{14.21}$$

we may write (14.20) also as

$$n^{opt} = \sqrt{\frac{k_1 s_x}{k_2 s_y} q}. \tag{14.22}$$

Although (14.22) has been specifically derived for the homogeneous distribution, it usually also gives a good approximation of the optimal value of n if the distribution of the required sizes is the bivariate normal or log normal distribution, provided the value of B is relatively large with respect to $k_1 s_x$ and $k_2 s_y$.

A similar expression as (14.22) can be derived for the case that the homogeneous distribution in combination with the quadratic loss function is used. Assuming that B is chosen so large that r_1 and r_2 equal their maximally permittable values, the expected adaptation loss of an optimal size pattern equals,

$$E(L) = \frac{1}{2}k \left[E(y) \left(\frac{x_{n+1} - x_0}{n} \right) + E(x) \left(\frac{y_{m+1} - y_0}{m} \right) \right.$$
$$\left. + \frac{1}{2} \left(\frac{x_{n+1} - x_0}{n} \right) \left(\frac{y_{m+1} - y_0}{m} \right) \right] \tag{14.23}$$

so that,

$$n^{opt} = \sqrt{\frac{E(y)\,s_x}{E(x)\,s_y}}\,q = \sqrt{\frac{\nu_x}{\nu_y}}\,q \tag{14.24}$$

where ν_x and ν_y equal the variation coefficients of x and y respectively. Also this formula gives usually a good indication how to choose n and m for a given value of the product, q, if the bivariate normal or log normal distribution is used with a large value of the factor B/k.

14.4.2 The Bivariate Normal Distribution and the Linear Loss Function

In this section we will use the linear loss function which can be obtained from (13.1) by putting $\alpha_1 = \alpha_2 = \beta_1 = \beta_2 = 1$. According to (14.2) and (14.3) the partial derivatives with respect to $x_1, \ldots, x_n$ are

$$\frac{\partial E(L)}{\partial x_i} = k_1 \int_{x_{i-1}}^{x_i} \int_{y_0}^{y_m} w(x,y)\,dy\,dx$$
$$- k_1(x_{i+1} - x_i) \int_{y_0}^{y_m} w(x_i,y)\,dy \quad i = 1, \ldots, n-1, \tag{14.25}$$

$$\frac{\partial E(L)}{\partial x_n} = k_1 \int_{x_{n-1}}^{x_n} \int_{y_0}^{y_m} w(x,y)\,dy\,dx$$
$$+ k_2 \sum_{j=1}^{m} \int_{y_{j-1}}^{y_j} (y_j - y)\,w(x_n,y)\,dy - B \int_{y_0}^{y_m} w(x_n,y)\,dy. \tag{14.26}$$

Applying the transformations (13.6) and the substitutions,

$$v_i = (x_i - \mu_x)/\sigma_x \quad i = 0, \ldots, n, \tag{14.27}$$

$$u_j = (y_j - \mu_y)/\sigma_y \qquad j = 0, \ldots, m \tag{14.28}$$

to (14.25) and (14.26) and putting the result equal to zero we have the recursion formulae

$$\int_{v_{i-1}}^{v_i} \int_{u_0}^{u_m} g(v,u)\, du\, dv = (v_{i+1} - v_i) \int_{u_0}^{u_m} g(v_i,u)\, du$$
$$i = 1, \ldots, n-1, \tag{14.29}$$

$$k_1 \int_{v_{n-1}}^{v_n} \int_{u_0}^{u_m} g(v,u)\, du\, dv + k_2 \frac{\sigma_y}{\sigma_x} \sum_{j=1}^{m} \int_{u_{j-1}}^{u_j} (u_j - u)\, g(v_n,u)\, du$$

$$= \frac{B}{\sigma_x} \int_{u_0}^{u_m} g(v_n,u)\, du \tag{14.30}$$

where the function g is defined by (13.7). The remaining m recursion formulae are obtained in the same way from the derivatives with respect to $y_1, \ldots, y_m$. Note that if ρ equals zero the equations (14.29) reduce to those of the one dimensional case (see Sec. 6.3). This is not so for (14.30) since the second term on the left-hand side does not occur in the corresponding equation of (6.17). If, however, B is large as compared to $k_1\sigma_x$ and $k_2\sigma_y$ and $k_1\sigma_x$ is about equal to $k_2\sigma_y$ the influence of this term will become negligible, indicating that the v and u coordinates of the optimal size pattern will closely approximate the corresponding optimal one dimensional size patterns.

As is proved in Appendix B [see result (c)], the integral, $\int_{u_0}^{u_m} g(v_i,u)\, du$ in (14.29) equals

$$f(v_i)\, F((u_m - \rho v_i)/\sqrt{1 - \rho^2}). \tag{14.31}$$

Using (14.31) and the substitutions,

$$v_k^j = (v_k - \rho u_j)/\sqrt{1 - \rho^2} \qquad k = 0, \ldots, n, \ j = 0, \ldots, m, \tag{14.32}$$
$$u_k^i = (u_k - \rho v_i)/\sqrt{1 - \rho^2} \qquad k = 0, \ldots, m, \ i = 0, \ldots, n, \tag{14.33}$$

we may reduce the set of equations (14.29) to

$$\int_{v_{i-1}}^{v_i} \int_{u_0}^{u_m} g(v,u)\, du\, dv = (v_{i+1} - v_i)\, f(v_i)\, F(u_m^i) \qquad i = 1, \ldots, n-1. \tag{14.34}$$

Hence

$$v_{i+1} = v_i + \left(\int_{v_{i-1}}^{v_i} \int_{u_0}^{u_m} g(v,u)\, du\, dv \right) / f(v_i) F(u_m^i) \qquad i = 1, \ldots, n-1. \tag{14.35}$$

Similarly we have

$$u_{j+1} = u_j + \left(\int_{v_0}^{v_n} \int_{u_{j-1}}^{u_j} g(v,u)\,du\,dv \right) / f(u_j) F(v_n^j) \qquad j = 1, \ldots, m-1. \tag{14.36}$$

Given u_m and v_1 we can calculate $v_2, \ldots, v_n$ from (14.35). The standardization points $v_1, \ldots, v_n$ are located optimally, given $u_1, \ldots, u_m$, if v_{n-1} and v_n satisfy (14.30). Using the results (c) and (d) of Appendix B we may reduce this equation to

$$k_1 \int_{v_{n-1}}^{v_n} \int_{u_0}^{u_m} g(v,u)\,du\,dv + k_2 \frac{\sigma_y}{\sigma_x} \sqrt{1 - \rho^2}\, f(v_n) \left[\sum_{j=1}^{m} \{ u_j^n \, (F(u_j^n) \right.$$
$$\left. - F(u_{j-1}^n)) \} + f(u_m^n) \right] = \frac{B}{\sigma_x} f(v_n)\, F(u_m^n). \tag{14.37}$$

Dividing (14.37) by k_1 we see that the location of the optimal size pattern depends only on the factors,

$$T_x = B/k_1 \sigma_x, \qquad T_y = B/k_2 \sigma_y \tag{14.38}$$

and the parameters n, m and ρ. For the calculation of the expected adaptation loss we use (14.1) with $\alpha_1 = \alpha_2 = \beta_1 = \beta_2 = 1$. Applying the transformations (13.6), the substitutions (14.27), (14.28), (14.32), (14.33) and (14.38) to (14.1) and using the result (f) of Appendix B we may write the expected adaptation loss as

$$E(L) = B \left[\sum_{i=1}^{n} \sum_{j=1}^{m} \left\{ \left(\frac{v_i}{T_x} + \frac{u_j}{T_y} \right) \int_{v_{i-1}}^{v_i} \int_{u_{j-1}}^{u_j} g(v,u)\,du\,dv \right\} \right.$$
$$+ f(v_n)\, F(u_m^n) \left(\frac{1}{T_x} + \frac{\rho}{T_y} \right) + f(u_m)\, F(v_n^m) \left(\frac{1}{T_y} + \frac{\rho}{T_x} \right)$$
$$\left. + 1 - \int_{v_0}^{v_n} \int_{u_0}^{u_m} g(v,u)\,du\,dv \right] . \tag{14.39}$$

14.5 SOME EXAMPLES

To illustrate the characteristics of size patterns chosen with the grid restriction we will give two examples. Both examples are based on the model discussed in Sec. 14.4.2. In the first example the parameter values T_x, T_y, n and m are chosen equal to

$$T_x = T_y = 25, n = m = 3.$$

The optimal transformed size patterns have been calculated for varying values of ρ. These size patterns are inserted into Table 14.1 and plotted in Fig. 14.2 together with the 75% and the 95% contour lines of the bivariate standard normal distribution.

From Table 14.1 we see that the location of the optimal size pattern is nearly independent of ρ. Also the expected adaptation losses and the coverages are almost the same for the three cases. Furthermore, if we compare the u or v coordinates of the sizes with the optimal one dimensional size pattern consisting of 3 sizes and with B/kσ equal to 25 (see Appendix A, Table A.1) it appears that no significant differences exist. Moreover, multiplying the expected adaptation loss of the latter size pattern by two we obtain an amount of .067 which equals two of the three calculated expected adaptation losses in Table 14.1 and hardly differs from the third. This leads to the idea that the optimal size pattern might as well be obtained by solving the corresponding one dimensional models with B/kσ chosen equal to T_x and T_y, respectively. Numerical calculation of some other size patterns showed that this method works satisfactorily, except if T_x differs relatively much from T_y. In all cases, however, the one dimensional optimal size patterns are very well suited to be used as starting sets to initialize the iterative procedure discussed in Sec. 14.2.

Table 14.1. Some Optimal Size Patterns for Varying ρ with $T_x = T_y = 25$ and n = m = 3

i	$v_i = u_i$	Probabilities Service Areas		
$\rho = 0$, Expected Adaptation Loss/B = .067, Coverage = .997				
1	−.077	.062	.053	.018
2	1.104	.186	.157	.053
3	2.927	.220	.186	.062
$\rho = .4$, Expected Adaptation Loss/B = .067, Coverage = .997				
1	−.079	.030	.063	.040
2	1.099	.154	.178	.063
3	2.917	.285	.154	.030
$\rho = .8$, Expected Adaptation Loss/B = .066, Coverage = .996				
1	−.089	.003	.053	.081
2	1.079	.099	.243	.053
3	2.855	.362	.099	.003

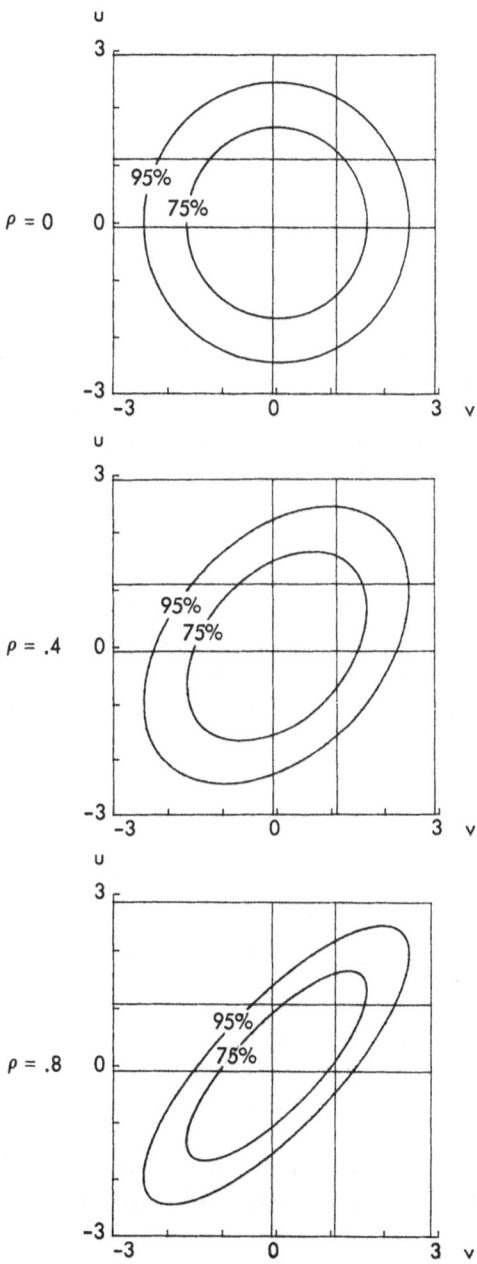

Figure 14.2. The Graphical Representation of the
Size Patterns Listed in Table 14.1

Next we will consider the efficiency of the grid restriction in the given example. This efficiency can be determined by calculating the (relative) increase of the expected adaptation loss caused by the removal of one or more sizes from the size pattern. In case many sizes can be deleted without influencing the expected loss very much we might say that the efficiency is low; if, on the other hand, none or only a few sizes can be deleted the efficiency is respectively high or reasonable.

To evaluate the efficiency of the grid restriction we have also inserted into Table 14.1 the probabilities of the service areas for the various cases. These probabilities are written in the form of a matrix; the (i,j)th element of this matrix equals the probability of the service area of size (v_j, u_{m+1-i}). As can be seen, the probabilities of the service areas of the sizes (v_1, u_3), (v_2, u_3), (v_3, u_3), (v_3, u_2) and (v_3, u_1) are relatively small for all values of ρ. If $\rho = 0$ these sizes cover only 25% of the sales which indicates that probably one or more of these sizes can be deleted from the size pattern without influencing the expected loss very much. Note that when deleting one or more sizes, the remaining sizes will not be located optimally anymore. The problem of how to find the optimal location of the remaining sizes is considered in Chap. 16. To determine the increase of the expected loss when deleting for instance size (v_1, u_3) we have to realize that the consumers whose required size was a point in the service area of (v_1, u_3) now are forced to purchase size (v_2, u_3). Without loss of generality B can be chosen equal to one so that according to (14.39) the expected adaptation loss increases with,

$$\{(v_2/T_x + u_3/T_y) - (v_1/T_x + u_3/T_y)\} \int_{v_0}^{v_1} \int_{u_2}^{u_3} g(v,u)\,du\,dv = .003$$

or about 4% of the calculated expected loss of .067. Using the same method we find that deleting (v_1, u_3) as well as (v_2, u_3) from the size pattern will cause an increase of .011 or 16%. Finally, deleting size (v_3, u_3) will increase the expected adaptation loss with .016 or 24%.

On the basis of these figures we conclude that the grid restriction gives a reasonably efficient solution of the standardization problem if $\rho = 0$. Only the benefits of the inclusion of size (v_1, u_3) and size (v_3, u_1) are questionable since the expected adaptation loss hardly increases if these sizes are deleted.

Turning now to the case $\rho = .8$ we see that the matrix of probabilities is entirely different from the matrix discussed above. The probability of the service area of size (v_1, u_3) now for instance equals .003 which means that this size will scarcely be purchased by the consumers. Deleting it raises the expected adaptation loss with .0001 or .1%. This obviously signifies that it is useless to include this size [and size (v_3, u_1)] in the size pattern. The efficiency of the grid restriction therefore must be considered to be low for larger positive values of ρ.

To illustrate the position of the optimal size pattern if T_x does not equal T_y the solutions of the recursion formulae have been calculated for the following parameter values:

$$T_x = 25, \qquad T_y = 10, 5 \text{ and } 1,$$
$$n = m = 3, \qquad \rho = 0.$$

The coordinates of the transformed sizes are inserted into Table 14.2 together with the probabilities of the service areas. The optimal transformed size patterns are plotted in Fig. 14.3 with the 95% contour lines of the bivariate standard normal distribution. If we put B = 25 and $\sigma_x = \sigma_y = 1$, it follows from the values of T_x and T_y that k_1 equals 1 and that k_2 respectively equals 2.5, 5 and 25. From Table 14.2 we see that, as could be expected, the optimal values of the standardization points u_1, u_2 and u_3 decrease for increasing k_2. The optimal values of v_1, v_2 and v_3 on the other hand remain about the same in the three cases. If we take the size pattern with $T_y = 10$ and evaluate its efficiency we arrive at the conclusion that the efficiency is lower than that of the corresponding problem with $T_x = T_y = 25$. This obviously is caused by the fact that k_2 has

Table 14.2. Some Optimal Size Patterns for Varying T_y with $\rho = 0$, $T_x = 25$ and n = m = 3

i	v_i	u_i	Probabilities Service Areas		
$T_y = 10$, *Expected Adaptation Loss*/B = .113, *Coverage* = .992					
1	−.080	−.154	.076	.064	.022
2	1.098	.959	.184	.155	.053
3	2.909	2.516	.205	.174	.059
$T_y = 5$, *Expected Adaptation Loss*/B = .183, *Coverage* = .982					
1	−.085	−.238	.090	.077	.026
2	1.088	.808	.179	.152	.052
3	2.880	2.145	.189	.161	.055
$T_y = 1$, *Expected Adaptation Loss*/B = .585, *Coverage* = .811					
1	−.122	−.658	.117	.102	.039
2	1.018	.136	.135	.118	.045
3	2.676	.892	.115	.101	.038

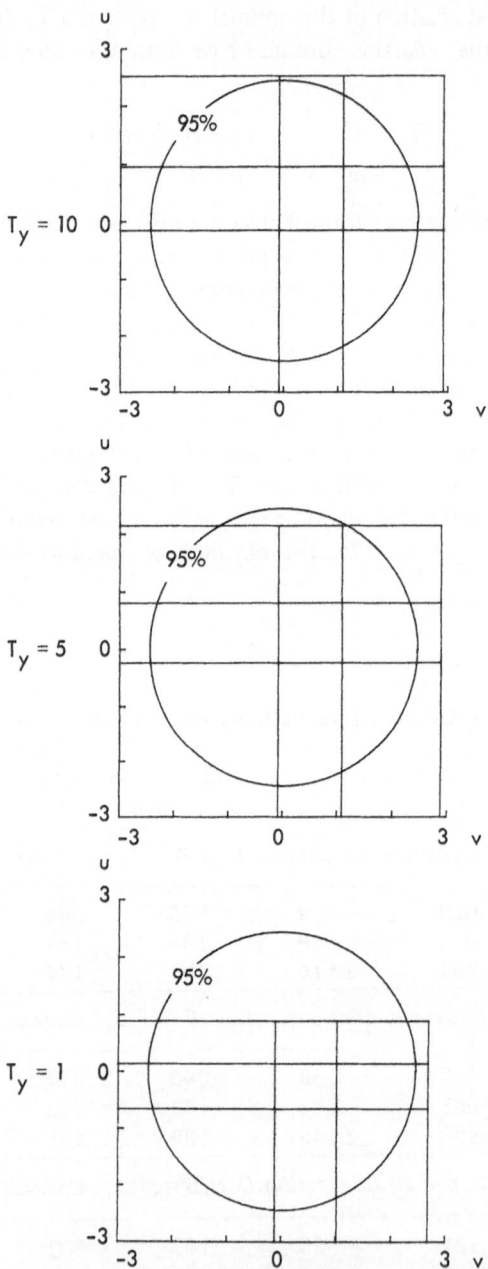

Figure 14.3. The Graphical Representation of the
Size Patterns Listed in Table 14.2

become a factor 2.5 larger than k_1 but that nevertheless as many standardization points are chosen with respect to the first property as to the second property. An obvious way of trying to reduce the expected adaptation loss and to improve the efficiency is therefore to choose more standardization points with respect to the second property. When keeping the number of sizes to be chosen about constant we then have two alternatives, namely to put n equal to 2 and m equal to 4 or to put n equal to 1 and m equal to 9. The optimal transformed size pattern with n = 2 and m = 4, which is inserted into Table 14.3, has an expected adaptation loss of .110. This value slightly differs from the value .113 in Table 14.2. On the other hand the size pattern has now—instead of 9 sizes—8 sizes which clearly indicates that it is more economical to use the size pattern with n = 2 and m = 4 than that with n = m = 3. Next, the optimal size pattern has been computed with n = 1, m = 9 and T_y = 10. The expected adaptation loss associated with this size pattern is .129 so that n = 2 and m = 4 remains the best choice. Note that according to (14.22), the optimal value of n is (approximately) equal to two also. Computing (14.22) with the parameter values of the last size pattern of Table 14.2 we find n = .6. The nearest integer value of n is unity which implies that m equals 9. Some calculations showed that a grid with n = 1 and m = 9 is indeed the optimal choice. The expected adaptation loss of the optimal size pattern with this grid structure equals .339 which is .25 or 42% less than the expected adaptation loss of the size pattern with n = m = 3.

Concluding we remark that the grid restriction possesses some advantages such as the simple structure of the size pattern and the ease with which the model can be solved. It will however be clear that the efficiency of the restriction heavily depends on the shape of the probability density function. Taking the density to be bivariate normal it appeared that the position of the size pattern was almost independent of ρ although the shape of the distribution

Table 14.3. The Optimal Size Pattern with ρ = 0, T_x = 25, T_y = 10, n = 2 and m = 4

i	v_i	u_i	Probabilities	Service Areas
$T_y = 10$,	Expected Adaptation Loss$/B = .110$,	Coverage $= .992$		
1	.559	-.506	.070	.028
2	2.646	.368	.181	.070
3	–	1.271	.240	.096
4	–	2.704	.218	.087

changes considerably for increasing ρ. Also the expected adaptation loss remained about the same for varying values of ρ. Since, if ρ increases, the area inside a given contour line becomes smaller, one would expect the distances between the sizes to decrease and as a consequence of this a decreasing expected adaptation loss. This, however, does not happen as a result of the structure of the size pattern enforced by the grid restriction. Therefore we shall introduce in the next chapter a restriction which can be used if ρ is large; this restriction will then lead to a more efficient solution than the grid restriction.

15 THE LINE RESTRICTION

15.1 INTRODUCTION

The line restriction is—as the grid restriction—a restriction on the relative position of the sizes. If the size pattern consists of n sizes and we denote the ith size by (x_i, y_i) $(i = 1, \ldots, n)$ then the line restriction orders

$$x_i > x_j \quad \text{and} \quad y_i > y_j \quad \text{if } i > j. \tag{15.1}$$

Taking for instance n equal to 3, a size pattern chosen according to the line restriction has the following form (see Fig. 15.1). The figure shows that the service areas, except S_1, all have a similar shape. As when using the grid restriction this has the effect that the recursion formulae can be derived easily and have a simple form.

The line restriction is—when using the bivariate normal or log normal distribution—clearly of interest if the correlation coefficient, ρ, has a large positive value. Because of the fact that the density of the distribution is then closely centered along a straight line or a curve, the use of the line restriction leads in that case to better solutions—in terms of the expected adaptation loss—than the use of the grid restriction. The applicability of the line restriction, however, is certainly not limited to cases where ρ is large. Suppose for instance that in the

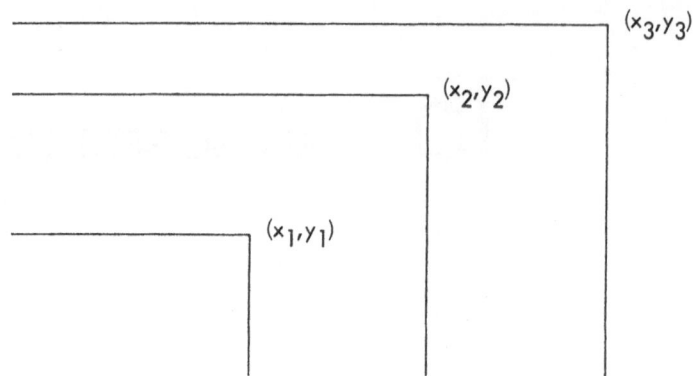

Figure 15.1.   An Example of the Line Restriction with n = 3

first example of Sec. 14.5 three sizes instead of nine sizes have to be chosen. When using the grid restriction we have two alternatives, which differ not essentially, namely to put n = 1 and m = 3 or to put n = 3 and m = 1. Both alternatives lead to an optimal transformed size pattern of which the sizes are located on a straight line, perpendicular to the v- or u-axis. The optimal transformed sizes obtained with the line restriction, however, are located on the 45° line through the origin. The latter size pattern obviously has—also for the case that $\rho = 0$—the smaller expected adaptation loss.

Though we consider the line restriction to be rather useful, it will not be treated very extensively. The reason is that the generalized grid restriction, which will be discussed in detail in the next chapter, can also be used to generate size patterns satisfying (15.1).

In the following section the recursion formulae are derived for the specific case that the linear loss function is used. Next we substitute in Sec. 15.3 the bivariate normal distribution into the recursion formulae. Two examples of the position of optimal size patterns are given in Sec. 15.4.

15.2 THE RECURSION FORMULAE FOR THE LINEAR
LOSS FUNCTION

The expected adaptation loss when using the linear loss function, defined by (13.1) with $\alpha_1 = \alpha_2 = \beta_1 = \beta_2 = 1$, can be written as,

$$E(L) = \sum_{i=1}^{n} \int_{S_i}\int \{k_1(x_i - x) + k_2(y_i - y)\}h(x,y)\,dy\,dx$$

$$+ B \left(1 - \sum_{i=1}^{n} \int_{S_i} \int h(x,y)\,dy\,dx \right)$$

$$= \sum_{i=1}^{n} \{(k_1 x_i + k_2 y_i)p_i\} - \int_{x_0}^{x_n} \int_{y_0}^{y_n} (k_1 x + k_2 y)h(x,y)\,dy\,dx$$

$$+ B \left(1 - \int_{x_0}^{x_n} \int_{y_0}^{y_n} h(x,y)\,dy\,dx \right) \tag{15.2}$$

where

$$p_i = \int_{x_0}^{x_i} \int_{y_0}^{y_i} h(x,y)\,dy\,dx - \int_{x_0}^{x_{i-1}} \int_{y_0}^{y_{i-1}} h(x,y)\,dy\,dx$$
$$i = 1, \ldots, n. \tag{15.3}$$

The derivatives with respect to $x_1, \ldots, x_n$ are

$$\frac{\partial E(L)}{\partial x_i} = k_1 p_i - \{k_1(x_{i+1} - x_i)$$

$$+ k_2(y_{i+1} - y_i)\} \int_{y_0}^{y_i} h(x_i,y)\,dy \quad i = 1, \ldots, n-1, \tag{15.4}$$

$$\frac{\partial E(L)}{\partial x_n} = k_1 p_n + k_2 \int_{y_0}^{y_n} (y_n - y)\, h(x_n,y)\,dy - B \int_{y_0}^{y_n} h(x_n,y)\,dy. \tag{15.5}$$

Putting (15.4), (15.5) and the derivatives with respect to $y_1, \ldots, y_n$ equal to zero we obtain the recursion formulae. These equations can be solved with the iterative procedure mentioned in Sec. 14.2.

15.3 THE BIVARIATE NORMAL DISTRIBUTION AND THE LINEAR LOSS FUNCTION

If the function w, defined by (13.5) is substituted for h in the recursion formulae, obtained from (15.4) and (15.5), we can simplify these formulae considerably by making use of the results (c) and (e) in Appendix B. Applying also the second transformation of (13.6) and the substitutions[1] (14.27), (14.28), (14.32), (14.33) and (14.38) we have

$$v_{i+1} = \frac{p_i}{f(v_i)\, F(u_i^i)} + v_i - \frac{T_x}{T_y}(u_{i+1} - u_i) \quad i = 1, \ldots, n-1, \tag{15.6}$$

[1] Note that as a consequence of the use of the line restriction n equals m.

$$P_n + \frac{T_x}{T_y} \sqrt{1 - \rho^2} f(v_n) \{u_n^n F(u_n^n) + f(u_n^n)\} = T_x f(v_n) F(u_n^n). \qquad (15.7)$$

The remaining n recursion formulae can be obtained by replacing in (15.6) and (15.7) T_x by T_y, T_y by T_x, v by u and u by v. From (15.6) and (15.7) we conclude in the first place that $v_i = u_i$ ($i = 1, \ldots, n$) if T_x equals T_y. In that case the optimal transformed size pattern thus can be found by means of the equations (15.6) and (15.7) only. Furthermore, if $T_x = T_y$ and ρ approaches one it follows that u_i^i ($= v_i^i$) ($i = 1, \ldots, n$) approaches zero and that p_i approaches $F(v_i) - F(v_{i-1})$ ($i = 1, \ldots, n$). The equations (15.6) then reduce to,

$$v_{i+1} = \frac{F(v_i) - F(v_{i-1})}{f(v_i)/2} + v_i - (v_{i+1} - v_i)$$

or

$$v_{i+1} = v_i + \frac{F(v_i) - F(v_{i-1})}{f(v_i)} \qquad i = 1, \ldots, n - 1. \qquad (15.8)$$

Because the second term of (15.7) approaches zero if ρ is close to one we may approximate (15.7) with

$$p_n = \frac{1}{2} T_x f(v_n). \qquad (15.9)$$

The set of equations (15.8) is equal to the first n - 1 recursion formulae of (6.17), derived in the one dimensional case. When comparing the nth equation of (6.17) with (15.9) we see that in (15.9) the factor T_x ($= B/k\sigma_x$) is divided by two. The cause of this difference is that the cost of the adaptation operation will on the average be twice as high in the two dimensional case because adaptation takes place with respect to two properties.

The second remark we want to make is a more general one and concerns two requirements any standardization model should satisfy. The first requirement is that the expected adaptation loss should be a decreasing function of the number of sizes to be chosen and the second is that the expected adaptation loss should approach zero if the number of sizes approaches infinity. Clearly, the model based on the grid restriction satisfies both requirements. When using the line restriction, however, the second requirement is violated. To make this point clear let us suppose that $T = T_x = T_y$. The optimal transformed sizes are then placed on the 45° line through the origin. Choosing an infinite number of sizes implies that every point on this line is a standardized size. The adaptation loss of a consumer with required size (v,u) is then, as easily can be verified, $(B/T) |v - u|$ and this term equals zero only if the required size is a point on the 45° line. The expected adaptation loss can, in this situation, be written as

$$\frac{B}{T}\int_{-\infty}^{\infty}\int_{-\infty}^{\infty}|v-u|g(v,u)du\,dv. \qquad (15.10)$$

As is proved in Appendix B, part (h), expression (15.10) can be reduced to

$$2B\sqrt{1-\rho}/T\sqrt{\pi}. \qquad (15.11)$$

The smallest expected adaptation loss that can be realized when using the line restriction may differ thus—especially for small values of ρ—considerably from zero. At first sight this seems to indicate that the line restriction has not much practical value. It appeared, however, that in quite some cases, the use of the line restriction becomes disadvantageous as compared with the use of the grid restriction only if a rather large number of sizes is to be chosen. Therefore, we are of the opinion that the line restriction might very well be useful when standardizing two properties of a product though one always has to take into account the effect described above.

15.4 SOME EXAMPLES

By means of two examples we will illustrate the model described in the previous section. In the first example the optimal transformed size patterns have been calculated for the parameter values,

$$T_x = T_y = 25, \qquad \rho = 0, .8$$

and for n (the number of sizes to be chosen) varying from 1 to 6. These size patterns are inserted into Table 15.1 with the probabilities of the service areas. Table 15.2 contains the expected adaptation loss and the coverage of each size pattern.

As can be seen from Table 15.1, the size patterns computed with $\rho = 0$ differ considerably from those computed with $\rho = .8$, whereas, on the other hand, the probabilities of the service areas hardly differ from each other for corresponding values of n. These results are just opposite to those which were obtained with the grid restriction. Recall that for the latter restriction, the position of the optimal size pattern was nearly independent of ρ, the probabilities of corresponding service areas differed substantially for varying ρ and some of the probabilities of the service areas approached zero for larger values of ρ. To compare the results obtained with the grid restriction with those of the line restriction we use Table 14.1 and Table 15.2 and put B equal to one. For the case that ρ equals .8, we read from Table 14.1 that—when employing the grid restriction with 9 sizes to

Table 15.1. Some Optimal Size Patterns for Varying n, with ρ equal to 0 and .8 Respectively and with $T_x = T_y = 25$

n	i	$v_i = u_i$	p_i	i	$v_i = u_i$	p_i
		$\rho = 0$			$\rho = .8$	
n = 1	1	2.11	.97	1	2.01	.97
n = 2	1	.96	.69	1	.70	.68
	2	2.60	.30	2	2.54	.31
n = 3	1	.47	.46	1	.13	.45
	2	1.42	.39	2	1.23	.39
	3	2.87	.15	3	2.81	.15
n = 4	1	.18	.33	1	-.22	.31
	2	.90	.34	2	.63	.34
	3	1.69	.24	3	1.53	.25
	4	3.04	.09	4	3.00	.09
n = 5	1	-.02	.24	1	-.46	.23
	2	.59	.28	2	.27	.28
	3	1.17	.25	3	.94	.25
	4	1.89	.17	4	1.74	.18
	5	3.17	.06	5	3.13	.06
n = 6	1	-.17	.19	1	-.65	.18
	2	.37	.23	2	.01	.22
	3	.85	.23	3	.57	.23
	4	1.37	.19	4	1.16	.20
	5	2.03	.12	5	1.90	.13
	6	3.28	.04	6	3.24	.04

Table 15.2. The Expected Adaptation Loss (Divided by B) and the Coverage of the Size Patterns Listed in Table 15.1

n	E(L)	Coverage	E(L)	Coverage
	$\rho = 0$		$\rho = .8$	
1	.201	.966	.196	.965
2	.126	.991	.112	.991
3	.099	.996	.083	.996
4	.086	.998	.067	.998
5	.078	.998	.058	.998
6	.073	.999	.052	.999

be chosen — an expected adaptation loss of .066 has been realized. For the same value of ρ, the use of the line restriction, however, leads to an expected adaptation loss of .067 if only 4 sizes are chosen. The performance of the grid restriction as compared with that of the line restriction is thus rather poor in this case. If ρ equals zero, the expected adaptation loss of the size pattern chosen with the grid restriction is .067. Applying the line restriction with 9 sizes we, however, find an expected adaptation loss of .063 indicating that even if ρ equals zero it might be advantageous to use the line restriction. When choosing more than 9 sizes, the results of the grid restriction soon become better than those of the line restriction. If, for instance, n equals 16 the expected adaptation loss of a size pattern selected according to the grid restriction equals .050 while the use of the line restriction leads to an expected adaptation loss of .055. Letting the number of sizes increase further, the expected adaptation loss approaches — when employing the line restriction — the limiting value .045 given by (15.11).

To determine the effect of a decreasing value of T_y on the position of the optimal size pattern, the solution of the recursion formulae has been calculated for T_y equal to 10 and 5 respectively. The other parameter values are

$$T_x = 25, \quad \rho = .8, \quad n = 6.$$

Table 15.3 contains the optimal size patterns with the associated expected adaptation losses. Furthermore, the size patterns are displayed in Fig. 15.2. As could be expected, the u coordinates of the sizes become smaller if T_y decreases. On the other hand, we see that the v coordinates — except v_6 — increase which has the effect that the corresponding probabilities of the service areas of the two size patterns remain about equal to each other.

Summarizing, we mention that the line restriction gives good results if the correlation coefficient, ρ, has a large positive value. Furthermore, if the number

Table 15.3. The Optimal Size Patterns for T_y Equal to 10 and 5 Respectively and with $\rho = .8$ and $T_x = 25$.

	$T_y = 10$, E(L)/B = .085, Coverage = .998				$T_y = 5$, E(L)/B = .130, Coverage = .994		
i	v_i	u_i	p_i	i	v_i	u_i	p_i
1	−.42	−.83	.17	1	−.22	−.92	.16
2	.21	−.17	.22	2	.38	−.26	.21
3	.75	.40	.22	3	.88	.29	.22
4	1.30	.98	.20	4	1.40	.85	.20
5	1.98	1.69	.13	5	2.02	1.53	.14
6	3.13	2.90	.05	6	2.99	2.58	.06

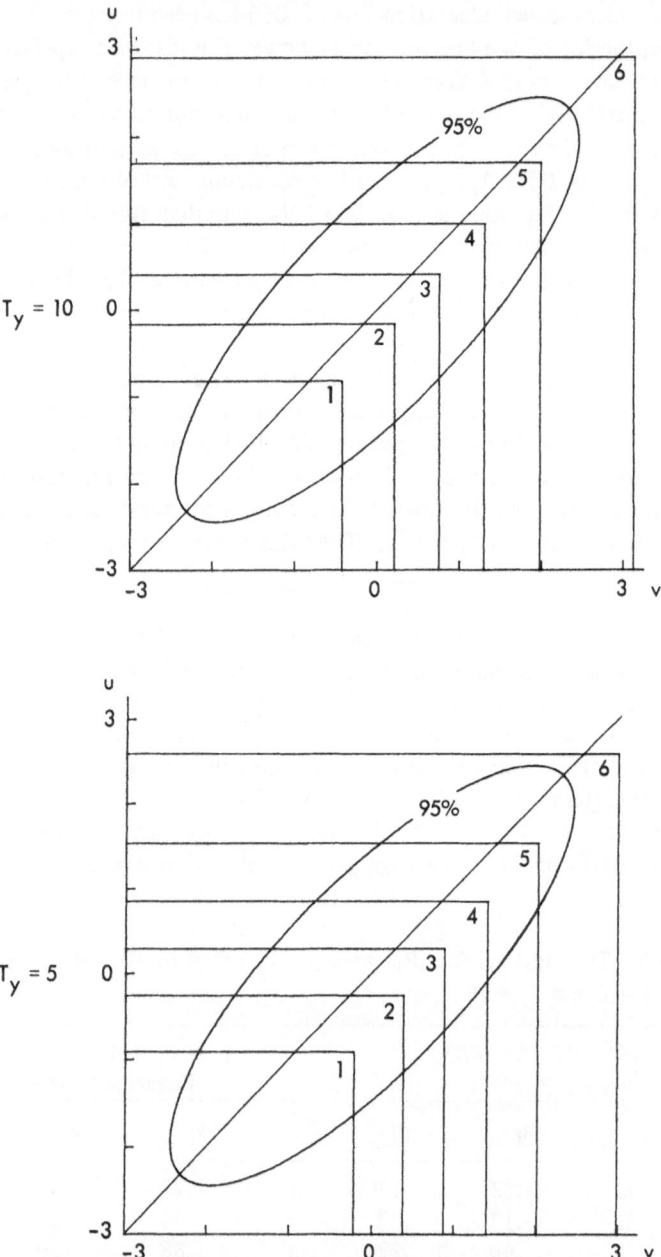

Figure 15.2. The Graphical Representation of the Size Patterns Listed
in Table 15.3

of sizes to be chosen is not too large, it might also be applied when ρ is zero or close to zero. An important point to notice is, finally, that we based our comparison between the performances of the line- and grid restriction on the expected adaptation loss. However, properly speaking, we should also take into account the production and distribution cost when making such a comparison. The size coordinates of the sizes chosen with the line restriction namely all may be different which is usually not the case when applying the grid restriction. This might have the effect that the total production and distribution costs are larger when using the line restriction since it seems reasonable to assume that these costs not only depend on the total number of sizes chosen but also on the number of different size coordinates.

16 THE GENERALIZED GRID
RESTRICTION

16.1 INTRODUCTION

The models that have been discussed in the previous two chapters are based on a rather simple structure of the service areas. Though they might be of use in some cases, it will be clear that in general the expected adaptation loss of size patterns computed by means of these models can be reduced by loosening or dropping the restriction the size pattern has to satisfy. Unrestricted selection of the sizes obviously leads to size patterns possessing the smallest expected loss. On the other hand, however, such size patterns might be unattractive from the practical point of view; as when using the line restriction the size coordinates will namely usually all be different. Consequently, the size pattern is "not easy to remember" but perhaps of more importance is the fact that—as already mentioned in Chap. 15—the production and distribution cost might be larger than those of a size pattern consisting of the same number of sizes but with a smaller number of different size coordinates.

Taking these considerations into account we have designed two—somewhat more elaborate—models. The first model, to be discussed in this chapter, is based on a restriction which is a generalization of the grid restriction. The restriction is therefore called "the generalized grid restriction." As we will see,

162

the generalized grid restriction is also a restriction on the relative position of the sizes but as compared with the grid restriction it enables us to take the characteristics of the probability density function of the required sizes much better into account when choosing the structure of the grid. In the next chapter we will pay attention to unrestricted selection of the sizes. A model will be constructed that can be used to generate size patterns having small expected adaptation losses. In general, when using this model, one will succeed in finding a size pattern with a smaller expected adaptation loss than that of a size pattern chosen with the generalized grid restriction. If, however, the number of sizes to be chosen becomes large, the relative as well as the absolute difference between the expected adaptation losses of the size patterns generated with the two models will be negligible. For the reasons mentioned above, we prefer in that case the use of the generalized grid restriction.

16.2 DESCRIPTION OF THE RESTRICTION

When employing the generalized grid restriction, it is—as with the grid restriction—supposed that n standardization points have to be chosen with respect to the first property (x) and m with respect to the second property (y). Contrary to the grid restriction, however, not every point (x_i, y_j) $(i = 1, \ldots, n, j = 1, \ldots, m)$ is necessarily a size. A point (x_i, y_j) is defined to be a size if and only if,

$$m_{1,i} \leqslant j \leqslant m_{2,i}, \quad i = 1, \ldots, n \qquad (16.1)$$

where the numbers $m_{k,i}$ $(k = 1, 2, i = 1, \ldots, n)$ have to be chosen such that they satisfy the conditions

$$
\begin{aligned}
&m_{1,i} \leqslant m_{2,i} \quad i = 1, \ldots, n, \\
&m_{1,1} = 1, \\
&m_{2,n} = m, \\
&m_{1,i+1} \leqslant m_{2,i} + 1 \quad i = 1, \ldots, n-1, \\
&m_{k,i} \leqslant m_{k,i+1} \quad k = 1, 2, \quad i = 1, \ldots, n-1.
\end{aligned}
\qquad (16.2)
$$

The first condition states that for every i there must be at least one j such that (x_i, y_j) is a size. The second and third conditions imply that the points in the left lower and the right upper corner of the grid, thus respectively (x_1, y_1) and (x_n, y_m) are always defined to be sizes. In combination with the fourth condition, these conditions assure that for every j there exists at least one i such that (x_i, y_j) is a size.

Summarizing we can say that the first four conditions guarantee that every standardization point is one of the coordinates of at least one size. To explain

Table 16.1.   The Values of $m_{k,i}$ (k = 1, 2, i = 1, ... , 6) of the Grid Displayed in Fig. 16.1

i	$m_{1,i}$	$m_{2,i}$
1	1	2
2	1	3
3	1	4
4	1	4
5	2	4
6	5	5

the meaning of the fifth condition a grid structure with n = 6 and m = 5 has been plotted in Fig. 16.1. The numbers $m_{k,i}$ are listed in Table 16.1.

The sizes are located at the intersection points of the horizontal and vertical lines. As can be seen, the three points (x_1,y_1), (x_2,y_1) and (x_3,y_1) are sizes. Suppose now, that (x_2,y_1) would not have been a size. This should have the consequence that a consumer whose required size is a point in the area denoted by the letter S, has to purchase one of the two sizes (x_2,y_2) or (x_3,y_1). Assuming the loss function and the values of the standardization points to be known, we can, for a given value of (x,y) determine which of the two sizes is bought by calculating the adaptation losses since it is supposed that the consumer purchases the size that leads to the smallest adaptation loss. If, however, the standardization points are not known, it is not possible to determine which size is bought and in that case we consequently do not know if (x,y) is a point in the service area of size (x_2,y_2) or of size (x_3,y_1). This has the effect that the expected adaptation loss and the recursion formulae cannot be written in an explicit form which makes the determination of the optimal size pattern considerably more difficult. As can be easily verified, the fifth condition prevents the occurrence of situations of the type described above; when applying the condition, the shapes of the service areas depend solely on the structure of the grid, thus on the numbers $m_{k,i}$, and not on the actual values of the standardization points.

16.3   THE RECURSION FORMULAE FOR LOSS FUNCTION I

When using the generalized grid restriction, the expected adaptation loss can be written in more than one form. We have chosen the form such that the derivatives with respect to $x_1, ... , x_n$ can be obtained in an easy way. This leads to,

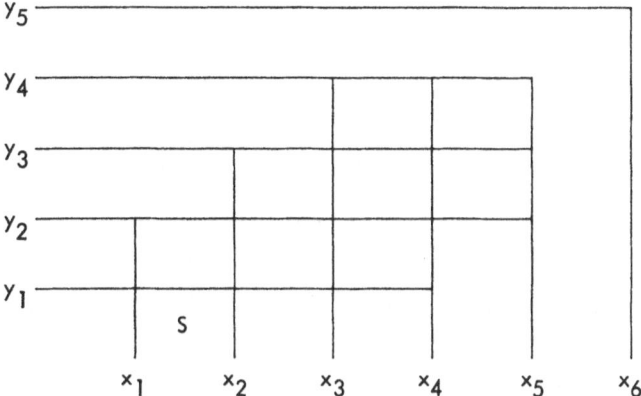

Figure 16.1. An Example of a Generalized Grid with n = 6 and m = 5

$$E(L) = \sum_{i=2}^{n} \sum_{j=m_{1,i}+1}^{m_{2,i-1}} \int_{x_{i-1}}^{x_i} \int_{y_{j-1}}^{y_j} \{k_1(x_i^{\alpha_1} - x^{\alpha_1})^{\beta_1}$$

$$+ k_2(y_j^{\alpha_2} - y^{\alpha_2})^{\beta_2}\}h(x,y)\,dy\,dx \tag{I}$$

$$+ \sum_{i=1}^{n} \int_{x_{i-1}}^{x_i} \int_{y_0}^{t_i} \{k_1(x_i^{\alpha_1} - x^{\alpha_1})^{\beta_1}$$

$$+ k_2(y_{m_{1,i}}^{\alpha_2} - y^{\alpha_2})^{\beta_2}\}h(x,y)\,dy\,dx \tag{II}$$

$$+ \sum_{i=1}^{n} \sum_{j=m_{2,i-1}+1}^{m_{2,i}} \int_{x_0}^{x_i} \int_{y_{j-1}}^{y_j} \{k_1(x_i^{\alpha_1} - x^{\alpha_1})^{\beta_1}$$

$$+ k_2(y_j^{\alpha_2} - y^{\alpha_2})^{\beta_2}\}h(x,y)\,dy\,dx \tag{III}$$

$$+ B\left(1 - \int_{x_0}^{x_n} \int_{y_0}^{y_m} h(x,y)\,dy\,dx\right) \tag{IV} \tag{16.3}$$

where

$$m_{2,0} = 1 \quad \text{and} \quad t_i = \min(y_{m_{1,i}}, y_{m_{2,i-1}}) \quad i = 1, \ldots, n.$$

Figure 16.2 shows how the integration process is performed. The numbers I, II and III in the grid correspond with those placed behind the various terms in (16.3). We see that the adaptation loss in the main body of the grid is calculated with the first term; thereafter follows the calculation of the adaptation loss in the lower and left edge of the grid by means of respectively the second and third term of (16.3). The fourth term, finally, gives the loss of not being served.

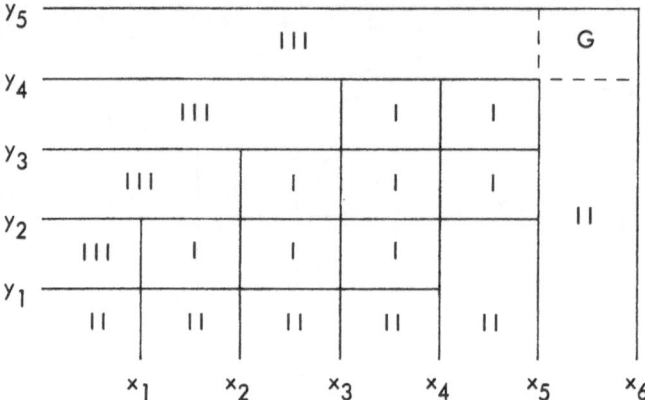

Figure 16.2. The Calculation of the Expected Adaptation Loss [see (16.3)]

From the figure we see also that the introduction of the variables t_i $(i = 1, \ldots, n)$ is necessary to avoid double counting. Replacing t_i by $y_{m_{1,i}}$ namely has the effect that the adaptation loss in the rectangle, denoted by the letter G, is calculated twice, once in the second term and once in the third term of (16.3). Differentating (16.3) with respect to the standardization points and putting the results equal to zero we obtain the recursion formulae. The derivatives with respect to $x_1, \ldots, x_n$ can, after some rearranging, be written as follows

$$
\begin{aligned}
\frac{\partial E(L)}{\partial x_i} = k_1 \beta_1 \alpha_1 x_i^{\alpha_1 - 1} &\left[\int_{x_{i-1}}^{x_i} \int_{y_0}^{t_i} (x_i^{\alpha_1} - x^{\alpha_1})^{\beta_1 - 1} h(x,y) dy\, dx \right.\\
&+ \sum_{j=m_{1,i}+1}^{m_{2,i-1}} \int_{x_{i-1}}^{x_i} \int_{y_{j-1}}^{y_j} (x_i^{\alpha_1} - x^{\alpha_1})^{\beta_1 - 1} h(x,y) dy\, dx \\
&\left. + \sum_{j=m_{2,i-1}+1}^{m_{2,i}} \int_{x_0}^{x_i} \int_{y_{j-1}}^{y_j} (x_i^{\alpha_1} - x^{\alpha_1})^{\beta_1 - 1} h(x,y) dy\, dx \right] \\
+ k_2 &\left[\int_{y_0}^{y_{m_{1,i}}} (y_{m_{1,i}}^{\alpha_2} - y^{\alpha_2})^{\beta_2} h(x_i,y) dy \right. \\
&+ \sum_{j=m_{1,i}+1}^{m_{2,i}} \int_{y_{j-1}}^{y_j} (y_j^{\alpha_2} - y^{\alpha_2})^{\beta_2} h(x_i,y) dy
\end{aligned}
$$

$$- \int_{y_0}^{t_{i+1}} (y_{m_{1,i+1}}^{\alpha_2} - y^{\alpha_2})^{\beta_2} h(x_i,y)\,dy$$

$$- \sum_{j=m_{1,i+1}+1}^{m_{2,i}} \int_{y_{j-1}}^{y_j} (y_j^{\alpha_2} - y^{\alpha_2})^{\beta_2} h(x_i,y)\,dy \Bigg]$$

$$- k_1 (x_{i+1}^{\alpha_1} - x_i^{\alpha_1})^{\beta_1} \int_{y_0}^{y_{m_{2,i}}} h(x_i,y)\,dy \qquad i = 1,\dots,n-1, \tag{16.4}$$

$$\frac{\partial E(L)}{\partial x_n} = k_1 \beta_1 \alpha_1 x_n^{\alpha_1 - 1} \left[\int_{x_{n-1}}^{x_n} \int_{y_0}^{t_n} (x_n^{\alpha_1} - x^{\alpha_1})^{\beta_1 - 1} h(x,y)\,dy\,dx \right.$$

$$+ \sum_{j=m_{1,n}+1}^{m_{2,n-1}} \int_{x_{n-1}}^{x_n} \int_{y_{j-1}}^{y_j} (x_n^{\alpha_1} - x^{\alpha_1})^{\beta_1 - 1} h(x,y)\,dy\,dx$$

$$+ \sum_{j=m_{2,n-1}+1}^{m} \int_{x_0}^{x_n} \int_{y_{j-1}}^{y_j} (x_n^{\alpha_1} - x^{\alpha_1})^{\beta_1 - 1} h(x,y)\,dy\,dx \Bigg]$$

$$+ k_2 \left[\int_{y_0}^{y_{m_{1,n}}} (y_{m_{1,n}}^{\alpha_2} - y^{\alpha_2})^{\beta_2} h(x_n,y)\,dy \right.$$

$$+ \sum_{j=m_{1,n}+1}^{m} \int_{y_{j-1}}^{y_j} (y_j^{\alpha_2} - y^{\alpha_2})^{\beta_2} h(x_i,y)\,dy \Bigg]$$

$$- B \int_{y_0}^{y_m} h(x_n,y)\,dy. \tag{16.5}$$

The derivatives with respect to $y_1,\dots,y_m$ can easily be derived from (16.4) and (16.5). Note, however, that for the calculation of the latter derivatives we have to use a table with typical elements $n_{k,j}$ ($k = 1,2, j = 1,\dots,m$) instead of the table used in (16.1). The numbers $n_{k,j}$ can be read from a figure showing the grid or they can be deduced from the numbers $m_{k,i}$.

From (16.4) we see that x_{i+1} can be expressed as an explicit function of x_{i-1}, x_i and $y_1,\dots,y_m$. This means that for a given set $y_1,\dots,y_m$, the optimal values of $x_1,\dots,x_n$ again can be found by means of the method of interval bisection. For the cases we have considered, the procedure to locate the optimal size pattern, described in Sec. 14.2, did work satisfactorily. However, although convergence usually is reached rather fast, it may occur that—depending on the structure of the grid—many iterations are required to find the optimal size pattern.

Having considered the question of how to find the optimal size pattern for a given structure of the grid we come now to the problem of choosing the values of n and $m_{k,i}$ ($k = 1, 2, i = 1, \ldots, n$) if the number of sizes is given. From (16.1) it follows that the number of sizes in the grid equals

$$\sum_{i=1}^{n} (m_{2,i} - m_{1,i} + 1).\qquad(16.6)$$

Putting (16.6) equal to the number of sizes to be chosen we see that usually many combinations of n and the $m_{k,i}$ can be found satisfying this restriction and the conditions (16.2). We have not found an analytical method to determine the optimal values of n and $m_{k,i}$, but practical experience indicated that the following procedure works satisfactorily. First, an optimal size pattern is determined using the grid restriction, described in Chap. 14, with n and m selected such that nm—the number of sizes in the grid—is considerably larger than the number of sizes that has to be chosen. Next, those sizes, which if being removed cause only a relatively small increase of the expected adaptation loss, are collected and omitted from the size pattern. The remaining sizes are then chosen optimally, using the generalized grid restriction, and this process is repeated until the number of sizes in the grid equals the number of sizes that has to be chosen.

16.4 THE RECURSION FORMULAE FOR LOSS FUNCTION II

If loss function II, defined by (13.3), is used and k is put equal to unity the expected adaptation loss can be written as

$$E(L) = \sum_{i=2}^{n} \sum_{j=m_{1,i}+1}^{m_{2,i-1}} x_i^{\alpha_1} y_j^{\alpha_2} \int_{x_{i-1}}^{x_i} \int_{y_{j-1}}^{y_j} h(x,y)\,dy\,dx$$

$$+ \sum_{i=1}^{n} x_i^{\alpha_1} y_{m_{1,i}}^{\alpha_2} \int_{x_{i-1}}^{x_i} \int_{y_0}^{t_i} h(x,y)\,dy\,dx$$

$$+ \sum_{i=1}^{n} \sum_{j=m_{2,i-1}+1}^{m_{2,i}} x_i^{\alpha_1} y_j^{\alpha_2} \int_{x_0}^{x_i} \int_{y_{j-1}}^{y_j} h(x,y)\,dy\,dx$$

$$- \int_{x_0}^{x_n} \int_{y_0}^{y_m} x^{\alpha_1} y^{\alpha_2} h(x,y)\,dy\,dx$$

$$+ B\left(1 - \int_{x_0}^{x_n} \int_{y_0}^{y_m} h(x,y)\,dy\,dx\right).\qquad(16.7)$$

The derivatives with respect to $x_1, \ldots, x_n$ are

$$\frac{\partial E(L)}{\partial x_i} = \alpha_1 x_i^{\alpha_1-1} \left[y_{m_{1,i}}^{\alpha_2} \int_{x_{i-1}}^{x_i} \int_{y_0}^{t_i} h(x,y)\,dy\,dx \right.$$

$$+ \sum_{j=m_{1,i}+1}^{m_{2,i-1}} y_j^{\alpha_2} \int_{x_{i-1}}^{x_i} \int_{y_{j-1}}^{y_j} h(x,y)\,dy\,dx$$

$$\left. + \sum_{j=m_{2,i-1}+1}^{m_{2,i}} y_j^{\alpha_2} \int_{x_0}^{x_i} \int_{y_{j-1}}^{y_j} h(x,y)\,dy\,dx \right]$$

$$+ x_i^{\alpha_1} \left[y_{m_{1,i}}^{\alpha_2} \int_{y_0}^{y_{m_{1,i}}} h(x_i,y)\,dy \right.$$

$$\left. + \sum_{j=m_{1,i}+1}^{m_{2,i}} y_j^{\alpha_2} \int_{y_{j-1}}^{y_j} h(x_i,y)\,dy \right]$$

$$- x_{i+1}^{\alpha_1} \left[y_{m_{1,i+1}}^{\alpha_2} \int_{y_0}^{t_{i+1}} h(x_i,y)\,dy \right.$$

$$\left. + \sum_{j=m_{1,i+1}+1}^{m_{2,i}} y_j^{\alpha_2} \int_{y_{j-1}}^{y_j} h(x_i,y)\,dy \right] \qquad i = 1, \ldots, n-1, \quad (16.8)$$

$$\frac{\partial E(L)}{\partial x_n} = \alpha_1 x_n^{\alpha_1-1} \left[y_{m_{1,n}}^{\alpha_2} \int_{x_{n-1}}^{x_n} \int_{y_0}^{t_n} h(x,y)\,dy\,dx \right.$$

$$+ \sum_{j=m_{1,n}+1}^{m_{2,n-1}} y_j^{\alpha_2} \int_{x_{n-1}}^{x_n} \int_{y_{j-1}}^{y_j} h(x,y)\,dy\,dx$$

$$\left. + \sum_{j=m_{2,n-1}+1}^{m} y_j^{\alpha_2} \int_{x_0}^{x_n} \int_{y_{j-1}}^{y_j} h(x,y)\,dy\,dx \right]$$

$$+ x_n^{\alpha_1} \left[y_{m_{1,n}}^{\alpha_2} \int_{y_0}^{y_{m_{1,n}}} h(x_n,y)\,dy \right.$$

$$\left. + \sum_{j=m_{1,n}+1}^{m} y_j^{\alpha_2} \int_{y_{j-1}}^{y_j} h(x_n,y)\,dy \right]$$

$$- x_n^{\alpha_1} \int_{y_0}^{y_m} y^{\alpha_2} h(x_n,y)\,dy - B \int_{y_0}^{y_m} h(x_n,y)\,dy. \quad (16.9)$$

Putting (16.8), (16.9) and the derivatives with respect to $y_1, \ldots, y_m$ equal to zero we have the recursion formulae. These can be solved with the procedure described in Sec. 14.2.

The model derived in this section has been used in a practical standardization problem. This application is discussed in Chap. 18. For some numerical examples we refer to that chapter.

16.5 THE BIVARIATE NORMAL DISTRIBUTION AND THE LINEAR LOSS FUNCTION

In this section we will assume the loss function to be linear. The appropriate recursion formulae, which can be obtained from (16.4) and (16.5), then can be reduced to a rather simple form by applying the transformations (13.6), the substitutions (14.27), (14.28), (14.32), (14.33), (14.38) and by making use of the results (c), (d) and (e) of Appendix B. This gives,

$$
\begin{aligned}
v_{i+1} = v_i &+ \frac{1}{f(v_i)\,F(u^i_{m_{2,i}})} \left[\int_{v_{i-1}}^{v_i} \int_0^{u_{m_{2,i-1}}} g(v,u)\,du\,dv \right. \\
&+ \int_{v_0}^{v_i} \int_{u_{m_{2,i-1}}}^{u_{m_{2,i}}} g(v,u)\,du\,dv \\
&+ \frac{T_x}{T_y} f(v_i) \left[u_{m_{1,i}} F(u^i_{m_{1,i}}) + \sum_{j=m_{1,i}+1}^{m_{2,i}} u_j \{ F(u^i_j) - F(u^i_{j-1}) \} \right. \\
&\left. - u_{m_{1,i+1}} F(r^i_{i+1}) - \sum_{j=m_{1,i+1}+1}^{m_{2,i}} u_j \{ F(u^i_j) - F(u^i_{j-1}) \} \right]
\end{aligned}
$$

$$i = 1, \ldots, n-1 \tag{16.10}$$

and

$$
\begin{aligned}
\int_{v_{n-1}}^{v_n} \int_0^{u_{m_{2,n-1}}} g(v,u)\,du\,dv &+ \int_{v_0}^{v_n} \int_{u_{m_{2,n-1}}}^{u_m} g(v,u)\,du\,dv \\
= \frac{T_x}{T_y} f(v_n) \left[u_{m_{1,n}} F(u^n_{m_{1,n}}) \right. &+ \sum_{j=m_{1,n}+1}^{m} u_j \{ F(u^i_j) - F(u^i_{j-1}) \} \\
&\left. - \rho v_n F(u^n_m) + \sqrt{1-\rho^2}\, f(u^n_m) \right] - T_x\, f(v_n)\, F(u^n_m).
\end{aligned}
\tag{16.11}
$$

The variable r^i_{i+1} appearing in (16.10) is defined as,

$$r^i_{i+1} = (r_{i+1} - \rho v_i)/(\sqrt{1-\rho^2}) \quad i = 1, \ldots, n-1 \tag{16.12}$$

where

$$r_i = (t_i - \mu_y)/\sigma_y \quad i = 1, \ldots, n.$$

The expected adaptation loss can also be written in terms of integrals of the standard normal- and the bivariate standard normal distribution. Applying the same transformations and substitutions as those used above, we have

$$
\begin{aligned}
E(L) = B \Bigg[& \sum_{i=2}^{n} \sum_{j=m_{1,i}+1}^{m_{2,i}-1} \left(\frac{v_i}{T_x} + \frac{u_j}{T_y} \right) \int_{v_{i-1}}^{v_i} \int_{u_{j-1}}^{u_j} g(v,u)\,du\,dv \\
& + \sum_{i=1}^{n} \left(\frac{v_i}{T_x} + \frac{u_{m_{1,i}}}{T_y} \right) \int_{v_{i-1}}^{v_i} \int_{u_0}^{r_i} g(v,u)\,du\,dv \\
& + \sum_{i=1}^{n} \sum_{j=m_{2,i-1}+1}^{m_{2,i}} \left(\frac{v_i}{T_x} + \frac{u_j}{T_y} \right) \int_{v_0}^{v_i} \int_{u_{j-1}}^{u_j} g(v,u)\,du\,dv \\
& + f(v_n)\, F(u_m^n) \left(\frac{1}{T_x} + \frac{\rho}{T_y} \right) + f(u_m)\, F(v_n^m) \left(\frac{1}{T_y} + \frac{\rho}{T_x} \right) \\
& + 1 - \int_{v_0}^{v_n} \int_{u_0}^{u_m} g(v,u)\,du\,dv \Bigg].
\end{aligned}
\tag{16.13}
$$

16.6 AN EXAMPLE

The example we will give concerns the model discussed in the previous section. In this example it is assumed that 9 sizes have to be selected. The remaining parameter values are

$$T_x = T_y = 25 \quad \text{and} \quad \rho = 0.$$

In Sec. 14.5 the optimal (transformed) size pattern for these values of the parameters was determined by making use of the grid restriction. Recall that in that section we came to the conclusion that the grid restriction led to a reasonably efficient solution for the given parameter values; only the sizes (v_1, u_3) and (v_3, u_1) could be omitted from the size pattern without changing the expected adaptation loss greatly. Therefore, when trying to reduce the expected adaptation loss by applying the generalized grid restriction instead of the grid restriction, it seems reasonable to begin with the replacement of these two sizes. Two obvious possibilities are respectively: moving both sizes to the right upper corner of the grid or moving one size to the right upper corner and the other to the lower left corner. Figure 16.3 shows the optimal size patterns for these two grid

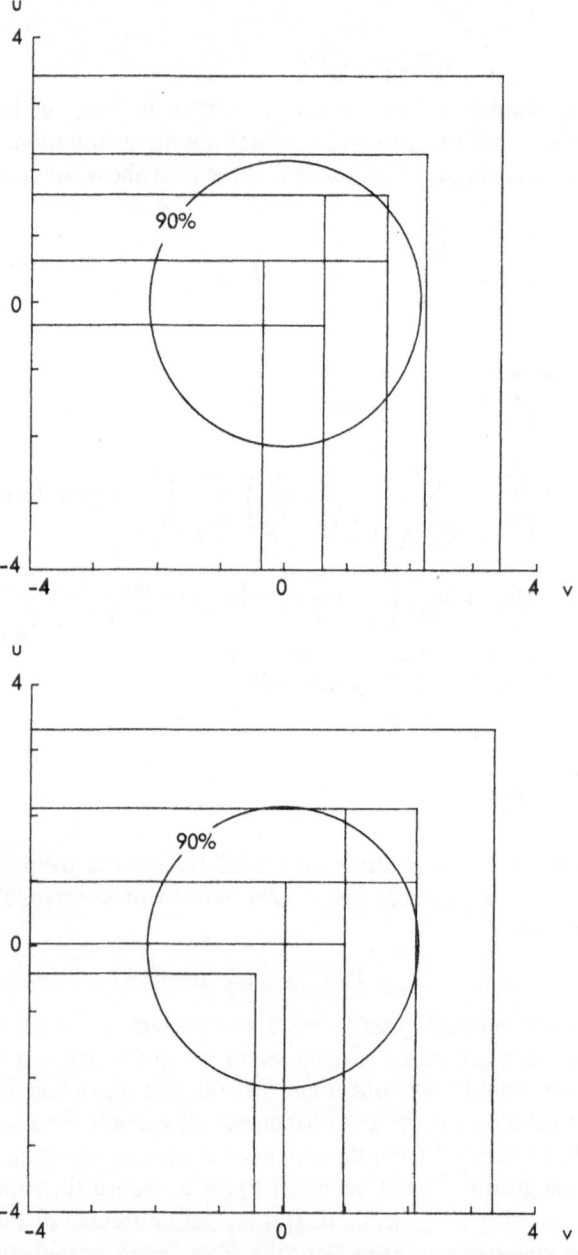

Figure 16.3. The Graphical Representation of the Size Patterns Listed in Table 16.2

structures. The values of the standardization points as well as the probabilities of the service areas are inserted in Table 16.2. The latter probabilities are again given in the form of a matrix where the (i,j)th element of this matrix denotes the probability of the service area of size (v_j, u_{m+1-i}). From the table we see that size pattern I has the smallest expected adaptation loss. It is .006 or 9 % less than the expected loss of the optimal size pattern chosen with the grid restriction (see Table 14.1). As is also done for the grid restriction, we can calculate the increase of the expected loss if one or more sizes are omitted from the size pattern. For size pattern I this leads to the conclusion that the removal of size (v_3, u_3) causes the smallest increase of the expected adaptation loss, namely .002 or 3.3%. Omitting one of the other sizes increases the expected loss with at least 8.4%. Though it can thus be said that the obtained grid offers an efficient solution for this problem, the small increase of the expected loss caused by the removal of size (v_3, u_3) indicates that perhaps still a better solution can be found. This proved indeed to be the case; by making use of the technique described in Sec. 16.3 we found a size pattern of which the expected loss was .0002 smaller than that of size pattern I. This size pattern is inserted into Table 16.3 and displayed in Fig. 16.4. The efficiency of the grid structure of size pattern III is higher than that of size pattern I. The removal of one of the sizes of the size pattern III namely causes an increase of the expected adaptation loss of at least 6%.

Table 16.2. Two Size Patterns, Consisting of 9 Sizes, with $T_x = T_y = 25$ and $\rho = 0$

i	$v_i = u_i$	Probabilities Service Areas				
Size Pattern I, Expected Adaptation Loss/B = .061, Coverage = .999						
1	−.348	−	−	−	−	.025
2	.621	−	−	−	.077	−
3	1.620	−	.157	.046	−	−
4	2.234	.134	.136	.157	−	−
5	3.423	.132	.134	−	−	−
Size Pattern II, Expected Adaptation Loss/B = .062, Coverage = .999						
1	−.427	−	−	−	−	.033
2	.032	−	−	.122	.021	−
3	.983	−	.166	.105	.122	−
4	2.113	−	.151	.166	−	−
5	3.334	.112	−	−	−	−

Table 16.3. The Optimal Size Pattern, Consisting of 9 Sizes, with $T_x = T_y = 25$ and $\rho = 0$

i	v_i	u_i	Probabilities Service Areas			
	Size Pattern III,	*Expected Adaptation Loss*/B = .061,	*Coverage = .999*			
1	−.167	−.341	–	–	–	.043
2	.936	.633	.103	.093	.037	–
3	2.089	1.945	.160	.145	.115	–
4	3.261	3.261	.159	.144	–	–

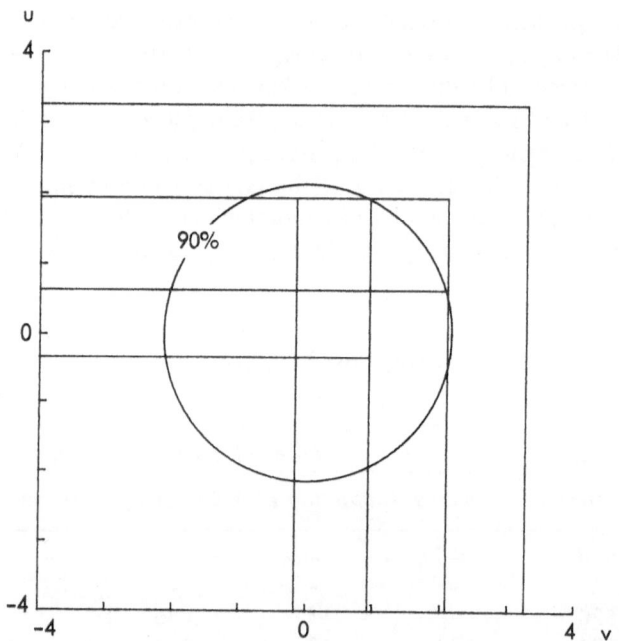

Figure 16.4. The Graphical Representation of the Size Pattern
Listed in Table 16.3

The very small difference between the expected adaptation losses of the two size patterns indicates that if a decision is made solely on the basis of the expected adaptation loss one will—in a practical situation—be indifferent between the two size patterns. Also taking into consideration the production cost, size pattern I might, however, very well be preferred to size pattern III as the latter size pattern has 7 different size coordinates while the first, being symmetrical with respect to the 45° line, has only 5 different size coordinates. By making use of the line restriction we find an expected adaptation loss of .063 if there are 9 sizes to be chosen. The use of the generalized grid restriction leads, relative to the use of the line restriction, to a decrease of the expected adaptation loss of .002 or 3.2% only. The decrease of the total cost might however, be considerably larger if we add the savings due to the choice of 5 instead of 9 different size coordinates.

16.7 AN EXTENSION OF THE MODEL

The conditions (16.1) and (16.2) have been introduced in order to obtain the recursion formulae in an easy way. As we saw, these formulae could be derived directly from the formula of the expected adaptation loss. If this requirement is dropped, we can construct a model which gives more freedom of choice with respect to the structure of the grid. Assuming that we still wish to avoid situations of the type described in the second part of Sec. 16.2, the grid might be constructed as follows. First, draw n lines perpendicular to the x-axis as is indicated by Fig. 16.5. Next, draw m horizontal lines perpendicular to the y-axis, satisfying the conditions that each horizontal line must at least intersect one

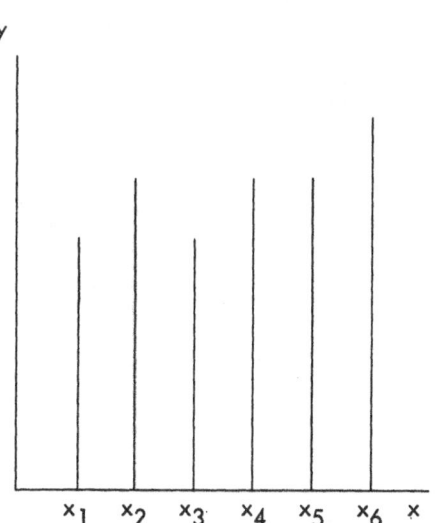

Figure 16.5. Another Method of Choosing the Grid (I)

vertical line and that each vertical line must at least once be intersected by a horizontal line. Placing the sizes at the intersection points of the horizontal and vertical lines we for instance arrive at the grid displayed in Fig. 16.6. Note, that—when constructing the grid in this way—the shapes of the serivce areas do not depend on the actual values of the standardization points. It will be clear that a general expression for the expected adaptation loss, such as for instance (16.3), valid for every possible grid structure does not exist. Therefore the formula of the expected adaptation loss becomes rather lengthy and differs from case to case which makes it less attractive to derive the recursion formulae from it. There is, however, another method to derive the recursion formulae. This method will be discussed extensively in the next chapter, where also a method for solving the recursion formulae is given. In our opinion there will be—after reading of Chap. 17—no new problems involved in finding a solution for the model described above which is the reason that we pay no further attention to it.

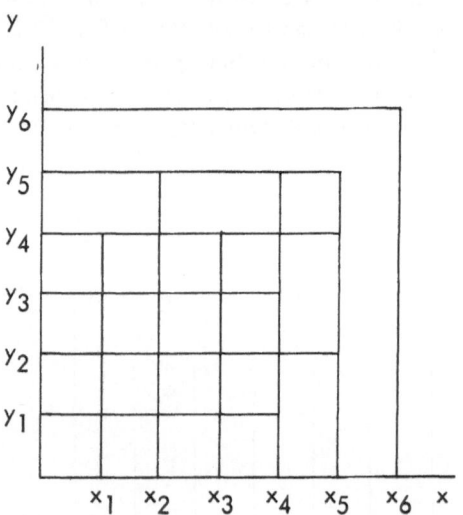

Figure 16.6. Another Method of Choosing the Grid (II)

17 UNRESTRICTED SELECTION OF THE SIZES

17.1 INTRODUCTION

The problems that arise if the sizes are to be chosen without a restriction on their relative position are far more complex than those we met in the previous chapters. This is mainly caused by the fact that the shape of the service areas is now in general not the same as was the case when using one of the restrictions. Since the shapes depend on the position of the sizes, the change of the position of one of the sizes might have the consequence that the shape of one or more of the service areas also changes considerably. This means that when an iterative method is used to solve the model the new shapes of the service areas have to be determined after each iteration. In order to complicate matters not too much, we will consider only those loss functions which lead to service areas whose boundaries consist of a number of horizontal and vertical line segments.

17.2 THE RECURSION FORMULAE FOR THE LINEAR LOSS FUNCTION

17.2.1   The Determination of the Service Areas

Since it is assumed that shift rule O is being applied, a consumer with required
size (x,y) has to choose a standardized size (x_i,y_i) with $x_i \geq x$ and $y_i \geq y$.
Furthermore, we suppose that he will select that size from the admissible standardized sizes, that leads to the smallest adaptation loss. In Fig. 17.1 a situation
is displayed where two standardized sizes, (x_i,y_i) and (x_j,y_j), are admissible to
a consumer with required size (x',y'). To determine which one he chooses we
draw a line through each size with the property that the size can be moved along
the line without changing the adaptation loss for the consumer. The equations
of the lines are

$$y = -\frac{k_1}{k_2}x + \frac{k_1}{k_2}x_i + y_i \quad i = 1, 2. \tag{17.1}$$

As can be easily verified, the length of the perpendicular b, which equals,

$$(k_1(x_i - x') + k_2(y_i - y'))/\sqrt{k_1^2 + k_2^2} \tag{17.2}$$

is proportional to the adaptation loss if size (x_i,y_i) is purchased. Hence, the
consumer chooses that size, that is located on the line with the shortest distance
to (x',y'). Using this rule the service areas can easily be determined if the size
pattern is given. In Fig. 17.2 the service areas are for example shown for a
size pattern consisting of 5 sizes with $k_1 = k_2$. The subscripts of the sizes are
chosen on the basis of the y coordinates of the sizes; size (x_1,y_1) is the size
with the smallest y coordinate, size (x_2,y_2) is the size with the smallest but one
y coordinate and so on.

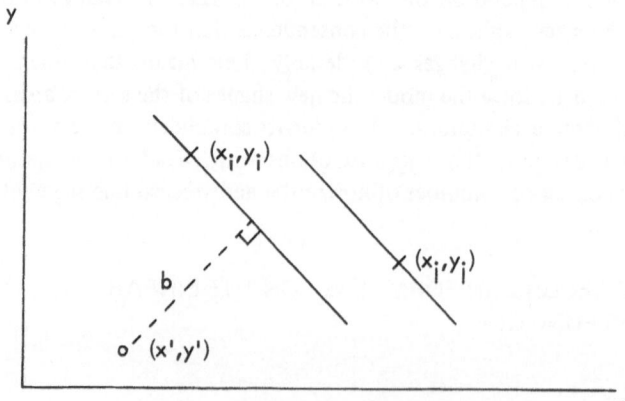

Figure 17.1.   The Choice between Two Standardized Sizes

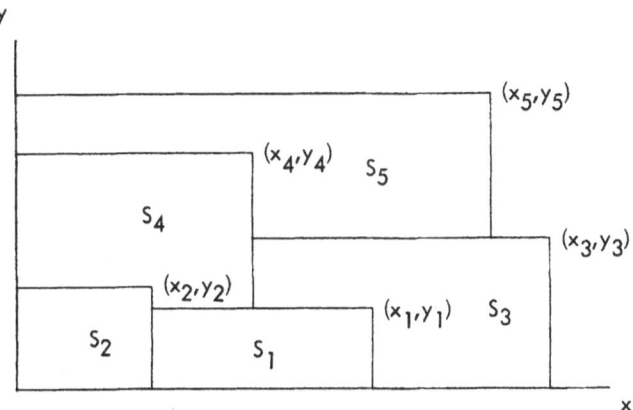

Figure 17.2. An Example of Unrestricted Selection of the Sizes
($n = 5$ and $k_1 = k_2$)

17.2.2 The Recursion Formulae and the Method of Solution

In its general form the expected adaptation loss can be written as

$$E(L) = \sum_{i=1}^{n} \int_{S_i} \int \{k_1(x_i - x) + k_2(y_i - y)\}h(x,y)\,dy\,dx$$

$$+ B \left(1 - \sum_{i=1}^{n} \int_{S_i} \int h(x,y)\,dy\,dx\right). \qquad (17.3)$$

Expression (17.3) shows that the derivatives of the expected adaptation loss
with respect to the size coordinates cannot be determined before the shapes of
the service areas are known. For a given size pattern, however, the derivatives
can be derived rather easily, since in that case the boundaries of the service areas
can be computed. To determine for example the derivative with respect to y_4, if
the sizes are located as displayed in Fig. 17.2, we first calculate the change of the
expected adaptation loss if a small amount, say δ, is added to y_4 (see Fig. 17.3).
The adaptation loss of the consumers whose required size is a point in S_4 will
increase by δk_2. On the other hand, the adaptation loss of the consumers whose
required size is a point in the shaded area of Fig. 17.3 will decrease by,

$$k_1(x_5 - x_4) + k_2(y_5 - (y_4 + \delta))$$

because these consumers now purchase size $(x_4, y_4 + \delta)$ instead of size (x_5, y_5).

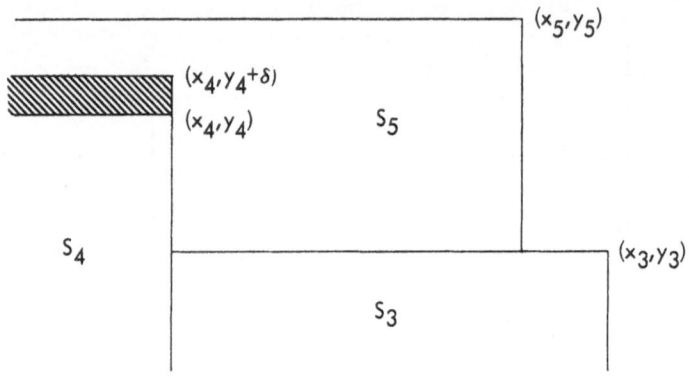

Figure 17.3. The Calculation of the Derivative of E(L) with Respect to y_4

Weighting these amounts with the relevant probabilities we have the change of the expected adaptation loss

$$\Delta E(L) = \delta k_2 \int_{S_4} \int h(x,y) \, dy \, dx - \{k_1(x_5 - x_4)$$

$$+ k_2(y_5 - (y_4 + \delta))\} \int_{x_0}^{x_4} \int_{y_4}^{y_4 + \delta} h(x,y) \, dy \, dx. \qquad (17.4)$$

Dividing (17.4) and δ and letting δ approach zero we find the first derivative of $E(L)$ with respect to y_4,

$$\frac{\partial E(L)}{\partial y_4} = k_2 \int_{S_4} \int h(x,y) \, dy \, dx - \{k_1(x_5 - x_4)$$

$$+ k_2(y_5 - y_4)\} \int_{x_0}^{x_4} h(x,y_4) \, dx. \qquad (17.5)$$

The derivative of $E(L)$ with respect to y_1 can be determined in the same way. This gives

$$\frac{\partial E(L)}{\partial y_1} = k_2 \int_{S_1} \int h(x,y) \, dy \, dx - \{k_1(x_3 - x_1) + k_2(y_3 - y_1)\} \int_{x_4}^{x_1} h(x,y_1) \, dx$$

$$- \{k_1(x_4 - x_1) + k_2(y_4 - y_1)\} \int_{x_2}^{x_4} h(x,y_1) \, dx. \qquad (17.6)$$

The derivative with respect to y_5 has a somewhat different form. As readily can be verified, this derivative namely equals

$$\frac{\partial E(L)}{\partial y_5} = k_2 \int\limits_{S_5} \int h(x,y)\,dy\,dx$$

$$+ k_1 \int_{x_0}^{x_5} (x_5 - x)h(x,y_5)\,dx - B \int_{x_0}^{x_5} h(x,y_5)\,dx. \quad (17.7)$$

Employing Newton's method to find a size pattern satisfying the recursion formulae we first compute for a given set of starting sizes, say z^i, the structure of the size pattern. For this purpose, an n by n matrix, say U, is constructed, which is derived in the following way (see Fig. 17.4). Since U contains all the necessary information about the shapes of the service areas, the vector of first derivatives,

$$W(z^i) = \left(\frac{\partial E(L)}{\partial x_1}, \ldots, \frac{\partial E(L)}{\partial x_n}, \frac{\partial E(L)}{\partial y_1}, \ldots, \frac{\partial E(L)}{\partial y_n} \right)^T \Bigg|_{z = z^i} \quad (17.8)$$

can be computed directly from z^i and U. To calculate the matrix of second derivatives, A, we use numerical differentiation. Note that usually not all n^2 second derivatives have to be computed. From Fig. 17.4 we see for instance that the first derivative of $E(L)$ with respect to x_2 does not contain x_3 so that $\partial E(L)/\partial x_2 \partial x_3$ and $\partial E(L)/\partial x_3 \partial x_2$ both equal zero. Using W and A, a new set of sizes z^{i+1} can be determined by means of (4.36). The whole procedure is then repeated until a size pattern satisfying the recursion formulae—within a given precision—is found. Although Newton's method proved to work extremely well in locating a size pattern satisfying the recursion formulae, several remarks have to be made. First, when using Newton's method, we will only succeed in finding a size pattern which satisfies the recursion formulae if the

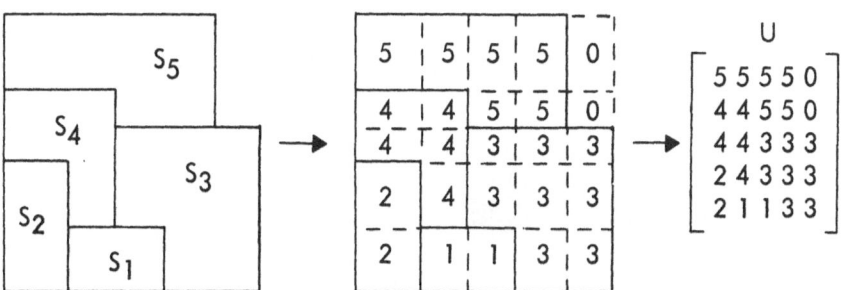

Figure 17.4. The Construction of the Matrix U

starting sizes are selected in such a way that the probabilities of the service areas are all of a reasonable magnitude. If the relative differences between these probabilities are too large, the method will diverge which means that after a number of iterations the probabilities of the service areas of some sizes become zero. Secondly, it is implicitly assumed by using first and second derivatives that the expected adaptation loss is differentiable everywhere. However, it can be shown that this is not the case. Consider for instance the Figs. 17.5, 17.6 and 17.7. In each of these figures two sizes are displayed together with the relevant part of their service areas. As can be seen, the x coordinate of size (x_1, y_1) increases in the three successive figures which causes a change in the shapes of the service areas. Consequently, it is not possible to determine what size the consumers, whose required size is a point in the shaded area of Fig. 17.6, will purchase. Both sizes, namely, cause the same adaptation loss. From the form of the derivative with respect to x_1 it then follows that there must exist a discontinuity in the first derivative at the point where x_1 equals the value indicated by Fig. 17.6. Hence, it may occur that during the iterative procedure size coordinates are generated some of whose first derivatives are nonexistent. The probability of the occurrence of such a situation, however, is almost zero, which is the reason that we neglected this problem. Another point needing attention is that a size pattern which satisfies the recursion formulae is in general not unique. Numerical experimentation showed that the number of solutions of the recursion formulae increases progressively with n which implies that for larger values of n not all solutions can be found. This means that it is not possible to find with certainty the size pattern with the smallest expected adaptation loss among the size patterns satisfying the recursion formulae since the latter are not all known. Our interest, however, is primarily aimed at the economic aspects of the problem and we are faced with the fact that in a practical situation a decision must be taken as to the choice of the sizes to be standardized. To provide a basis for such a decision we propose the calculation of a large number of different

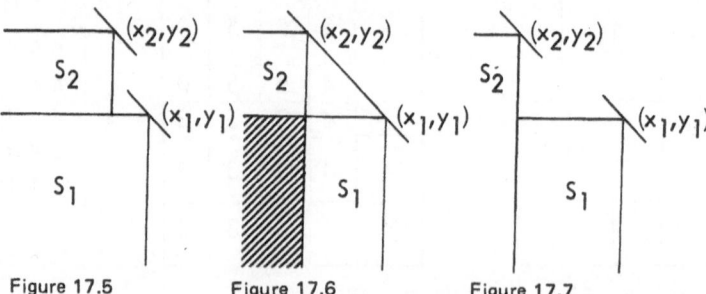

Figure 17.5   Figure 17.6   Figure 17.7

Figures 17.5–17.7. The Change of the Serivce Areas If x_1 Is Increased size patterns satisfying the recursion formulae. These size patterns can be obtained by choosing a large number of starting sets and using these as input for the iterative procedure mentioned above. The choice of the starting sets might be done in two ways.

In the first place the starting sets can be chosen by hand. It is then recommended to draw a number of contour lines of the distribution of the required sizes. Using these contour lines, one can roughly estimate the probabilities of the service areas of a given starting set so that eventually some sizes can be replaced if the probabilities of their service areas turn out to be too low. The next problem is how to select the starting sets efficiently. It will be clear that two starting sets which are about equal to each other shall in general lead to the same (sub)optimal size pattern. Our purpose, however, is to obtain as many (sub)optimal size patterns as possible from a given number of starting sets. It appeared that this can be done in a rather systematic way, namely by arranging the starting sets according to the permutation of the x coordinates. The permutation of the size pattern displayed in Fig. 17.2, for instance, can be determined in the following way. The smallest x coordinate belongs to size (x_2, y_2); therefore the first number of the permutation is 2. The one but smallest x coordinate is x_4 so that the next number of the permutation is 4. Proceeding in this way we find that the permutation of the size pattern equals $(2,4,1,5,3)$. When applying Newton's method, using successively two different starting sets with the same permutation, it became apparant that in most cases these starting sets converged to the same (sub)optimal size pattern with a permutation equal to that of the starting sets. On the other hand, two starting sets based upon a different permutation usually lead to different solutions. For a size pattern consisting of n sizes there exist n! permutations. However, it is certainly not necessary to try them all out. It, for instance, does not make any sense to select a starting set based on the permutation $(n, n-1, \ldots, 2, 1)$ if the distribution of the required sizes is bivariate normal with $\rho > 0$. Omitting the permutations that are not acceptable on intuitive grounds still leaves a large number of permutations that might be taken into consideration. Some experimentation with the model showed however that the sensitivity of the expected adaptation loss is rather low with respect to the permutation selected, provided that the starting sizes are reasonably scattered over the domain of the required sizes and that the probabilities of their service areas are not too close to zero. This implies that—although maybe the global optimum cannot be found with certainty—it is, using the described method, very well possible to find a number of acceptable (sub)optimal size patterns. Of these size patterns—all satisfying the recursion formulae and therefore at least possessing local minima of the expected adaptation loss—one can be chosen as the ultimate size pattern according to which the standardization takes place.

The second method that can be used to select the starting sets is to generate them randomly. Every size coordinate of a starting set is then obtained by putting it equal to a random variable drawn from the distribution of the required sizes. This method, of course, has the disadvantage that the starting sets often do contain sizes which are nearly equal to each other. Consequently, the service areas of some sizes might have a probability near zero so that, when applying Newton's method, no convergence is reached. On the other hand, if the procedure diverges, it is rather easy to draw a new starting set and to restart the iterations. For 8 sizes to be chosen, we found that about 20% of the starting sets led to a solution in case the computer program did not undertake any corrective action if the procedure diverged. By letting the program replace those sizes whose probabilities of the service areas became—during the iterations—too close to zero, this percentage increased to about 40%. The major advantage of selecting the sizes according to this method is that it works rather fast since the time consuming choice of the starting sets by hand is omitted. Furthermore, by generating a number of (sub)optimal size patterns with this method, we can get a general idea about the location of size patterns with a low expected adaptation loss. Next, we may try to reduce the expected adaptation loss of one or more of these size patterns further by replacing some sizes and continue the iterations.

17.3   AN EXAMPLE

To illustrate the model discussed in the previous section, five transformed size patterns—satisfying the recursion formulae—have been generated by making use of the bivariate normal distribution and the linear loss function. The starting sizes were chosen using a graph with a number of contour lines of the bivariate standard normal distribution. We only used starting sets (permutations) for which an acceptable solution could be expected. The parameter values and the five size patterns are listed in Table 17.1. Two of these size patterns are plotted in Fig. 17.8 together with the 95% contour line of the bivariate normal distribution. The table shows that size pattern V has the largest expected adaptation loss. This is caused by the fact that the sizes of this size pattern are located on the line $u = v$. The sizes of the remaining size patterns are more scattered over the domain of the required sizes, as can be seen from the matrices showing the structure of the size patterns. Furthermore, the table contains, besides size pattern V, two other size patterns which are symmetrical around the 45° line. These are size pattern II and size pattern IV. Interesting to note is that the size pattern with the smallest expected adaptation loss (i.e. size pattern I) does not possess this property.

The table shows also that in all cases most of the sizes are located in the first quadrant. We wish to emphasize that this is not an accidental result, obtained in this specific example. Some optimal size patterns computed with the quadratic

Table 17.1. Some (Sub)optimal Size Patterns, Consisting of 7 Sizes, with T_x = T_y = 25 and ρ = 0

Size Pattern	i	v_i	u_i	p_i	Structure Size Pattern							E(L)/B
I	1	-.19	-.19	.18	7	7	7	7	7	7	7	
	2	1.12	-.02	.17	6	6	6	6	6	7	7	
	3	.04	.67	.20	5	5	5	6	6	7	7	.065
	4	2.08	.68	.13	5	5	5	4	4	4	7	
	5	.54	1.42	.17	3	3	5	4	4	4	7	
	6	1.43	2.19	.10	3	3	2	2	4	4	7	
	7	3.24	3.24	.04	1	3	2	2	4	4	7	
II	1	-.16	-.16	.19	7	7	7	7	7	7	7	
	2	1.58	-.05	.14	6	6	6	6	6	6	7	
	3	.39	.39	.24	5	5	6	6	6	6	7	
	4	1.19	1.19	.13	5	5	4	4	6	6	7	.066
	5	-.05	1.58	.14	3	3	3	4	6	6	7	
	6	2.04	2.04	.12	3	3	3	2	2	6	7	
	7	3.28	3.28	.04	1	3	3	2	2	6	7	
III	1	-.25	-.25	.16	7	7	7	7	7	7	7	
	2	1.57	.16	.16	6	6	6	6	6	6	7	
	3	.27	.27	.21	5	5	5	5	6	6	7	
	4	.41	1.03	.19	4	4	4	5	6	6	7	.066
	5	1.03	1.62	.14	3	3	4	5	6	6	7	
	6	2.09	2.09	.11	3	3	4	2	2	6	7	
	7	3.32	3.32	.04	1	3	4	2	2	6	7	
IV	1	-.20	-.20	.18	7	7	7	7	7	7	7	
	2	2.07	.22	.11	6	6	7	7	7	7	7	
	3	.34	.34	.22	6	6	5	5	5	7	7	
	4	.80	.80	.22	4	4	4	4	5	7	7	.067
	5	1.73	1.73	.10	3	3	3	4	5	7	7	
	6	.22	2.07	.11	3	3	3	4	2	2	7	
	7	3.19	3.19	.05	1	3	3	4	2	2	7	
V	1	-.30	-.30	.15	7	7	7	7	7	7	7	
	2	.21	.21	.19	6	6	6	6	6	6	7	
	3	.63	.63	.20	5	5	5	5	5	6	7	
	4	1.05	1.05	.19	4	4	4	4	5	6	7	.069
	5	1.52	1.52	.15	3	3	3	4	5	6	7	
	6	2.15	2.15	.09	2	2	3	4	5	6	7	
	7	3.36	3.36	.03	1	2	3	4	5	6	7	

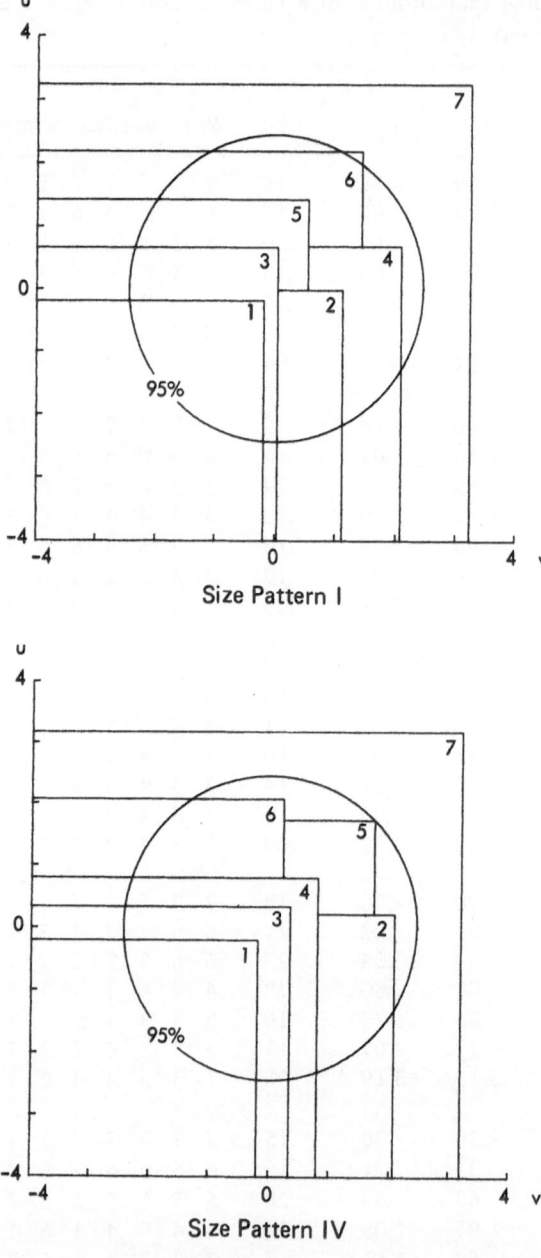

Figure 17.8. The Graphical Representation of Size Pattern
I and IV (See Table 17.1)

loss function and the bivariate normal distribution namely showed the same characteristic as soon as T_x and T_y were about equal to each other and of a reasonable magnitude. Furthermore, when applying the line restriction or the generalized grid restriction, most of the sizes usually are located in the first quadrant too (see for instance Table 15.1 and Table 16.3). The expected adaptation loss increases rather fast if the sizes are not preponderantly concentrated in the first quadrant. This became obvious after an experiment where several persons were asked to choose that size pattern that in their opinion would lead to the smallest expected adaptation loss. The expected adaptation losses of these size patterns were on the average 30% higher than the minimum expected adaptation loss (i.e. .065) we found. This was caused mainly by the fact that in most cases too many sizes of the intuitive size patterns were put in the other three quadrants.

17.4   THE RECURSION FORMULAE FOR LOSS FUNCTION II

17.4.1   The Determination of the Serivce Areas

If loss function II, defined by (13.3), is applied we may use the same principle as in Sec. 17.2.1 to find the service areas for a given set of sizes. Only the straight lines drawn through the sizes have to be replaced by curves of the form

$$y = y_i (x_i/x)^{\alpha_1/\alpha_2}. \qquad (17.9)$$

Because of the fact that shift rule O is to be applied, the boundaries of the service areas will again consist of horizontal and vertical line segments only.

17.4.2   The Recursion Formulae

In case loss function II is used the expected adaptation loss can be written as

$$E(L) = \sum_{i=1}^{n} \int_{S_i} \int k(x_i^{\alpha_1} y_i^{\alpha_2} - x^{\alpha_1} y^{\alpha_2}) h(x,y) \, dy \, dx$$

$$+ B \left(1 - \sum_{i=1}^{n} \int_{S_i} \int h(x,y) \, dy \, dx \right). \qquad (17.10)$$

To determine the derivatives with respect to the size coordinates we use the same technique as in Sec. 17.2.2. If we assume that the sizes are chosen as displayed in Fig. 17.2, we have for example,

$$\frac{\partial E(L)}{\partial y_4} = k\alpha_2 x_4^{\alpha_1} y_4^{\alpha_2-1} \int\int_{S_4} h(x,y)\,dy\,dx$$

$$- k(x_5^{\alpha_1} y_5^{\alpha_2} - x_4^{\alpha_1} y_4^{\alpha_2}) \int_{x_0}^{x_4} h(x,y_4)\,dx, \qquad (17.11)$$

$$\frac{\partial E(L)}{\partial y_5} = k\alpha_2 x_5^{\alpha_1} y_5^{\alpha_2-1} \int\int_{S_5} h(x,y)\,dy\,dx$$

$$+ ky_5^{\alpha_2} \int_{x_0}^{x_5} (x_5^{\alpha_1} - x^{\alpha_1})h(x,y_5)\,dx - B \int_{x_0}^{x_5} h(x,y_5)\,dx. \qquad (17.12)$$

To solve the recursion formulae the method of Newton can be used (see Secs. 4.6.3 and 17.2.2). An illustration of the model is given in the next chapter.

18 THE STANDARDIZATION OF FLAT STEEL SECTIONS

18.1 INTRODUCTION

The application to be discussed in this chapter concerns the standardization of flat steel sections for general use. The properties taken into consideration for standardization are the thickness (x) and the width (y), with $y \geqslant x$, of the sections. The data concerning the demand were gathered in 1962 from five large firms in the Netherlands. These data give the number of running meters of flat steel demanded for different combinations of the thickness (in mm) and width (in mm). Analysis of the data showed that a bivariate log normal distribution gives a very good fit to the observed frequencies [de Silva, (1969)].

The estimated values of the parameters of this distribution are,

$$\hat{\mu}_1 = 2.155, \quad \hat{\sigma}_1 = .403, \quad \hat{\rho} = .506,$$
$$\hat{\mu}_2 = 3.914, \quad \hat{\sigma}_2 = .513, \tag{18.1}$$

which leads to

$$E(x) = 9.36, \quad V(x) = 15.44, \quad \sqrt{V(x)} = 3.93,$$
$$E(y) = 57.14, \quad V(y) = 983.07, \quad \sqrt{V(y)} = 31.35. \tag{18.2}$$

189

When purchasing flat steel a consumer with required size (x,y) in general shall buy a size (x_i, y_i) so that $x \leqslant x_i$ and $y \leqslant y_i$. There may exist some exceptions, for instance if the flat steel is bought to resist a tensile force in which case the product $x_i y_i$ has to be larger than or equal to xy. We shall however neglect this complication and suppose that the consumers apply shift rule O with respect to both product properties. Furthermore, we assume that the adaptation loss consists of the loss of steel if the consumer buys a standardized size which differs from his required size. Since the price per kg of steel is nearly constant for flat steel with a thickness larger than 5 mm and a width larger than 19 mm, a region which covers more than 96% of the demand, we can calculate the economic loss by multiplying the adaptation loss in cubic mm with the product of the weight in kg per mm^3 and the (mean) price per kg. Denoting the latter product by k, we may write the adaptation loss that per running mm arises if size (x_i, y_i) is purchased whereas size (x,y) is required as

$$L(x,y) = k(x_i y_i - xy) \quad x \leqslant x_i \text{ and } y \leqslant y_i. \tag{18.3}$$

As the position of the size pattern does not depend on the individual values of k and B but only on the ratio B/k we have put k equal to one. Because no information about the loss of not being served was available B was chosen equal to 1000 in order to assure a reasonable coverage of the size pattern for larger values of n.

18.2 SOME ILLUSTRATIVE MODELS

To get a general idea of the shape of the (log normal) distribution of the required sizes, the 25%, 50%, 75%, 90%, 95% and 99% contour lines[1] are plotted in Fig. 18.1. We conclude from this figure that a size pattern based on the grid restriction may give good results in terms of expected adaptation loss. Using this restriction a number of optimal size patterns have been calculated for several values of n and m. These size patterns are listed in Table 18.1. This table also contains the expected adaptation losses and the probabilities of the service areas. Furthermore, the size patterns are plotted in the Figs. 18.2–18.7 together with the 95% contour line of the distribution of the required sizes.

From Table 18.1 we see that the rate of decrease of the expected adaptation loss per added size changes considerably if the number of sizes (nm) increases. If, for instance, the number of sizes is increased from 1 to 4 the expected adaptation loss decreases with 214, implying an average decrease per size of 71.3. On the other hand, if the number of sizes is increased from 36 to 64, the expected adaptation loss decreases with 44 so that the average decrease per size now only is 1.6.

---

[1] See Appendix C for the derivation of the contour lines.

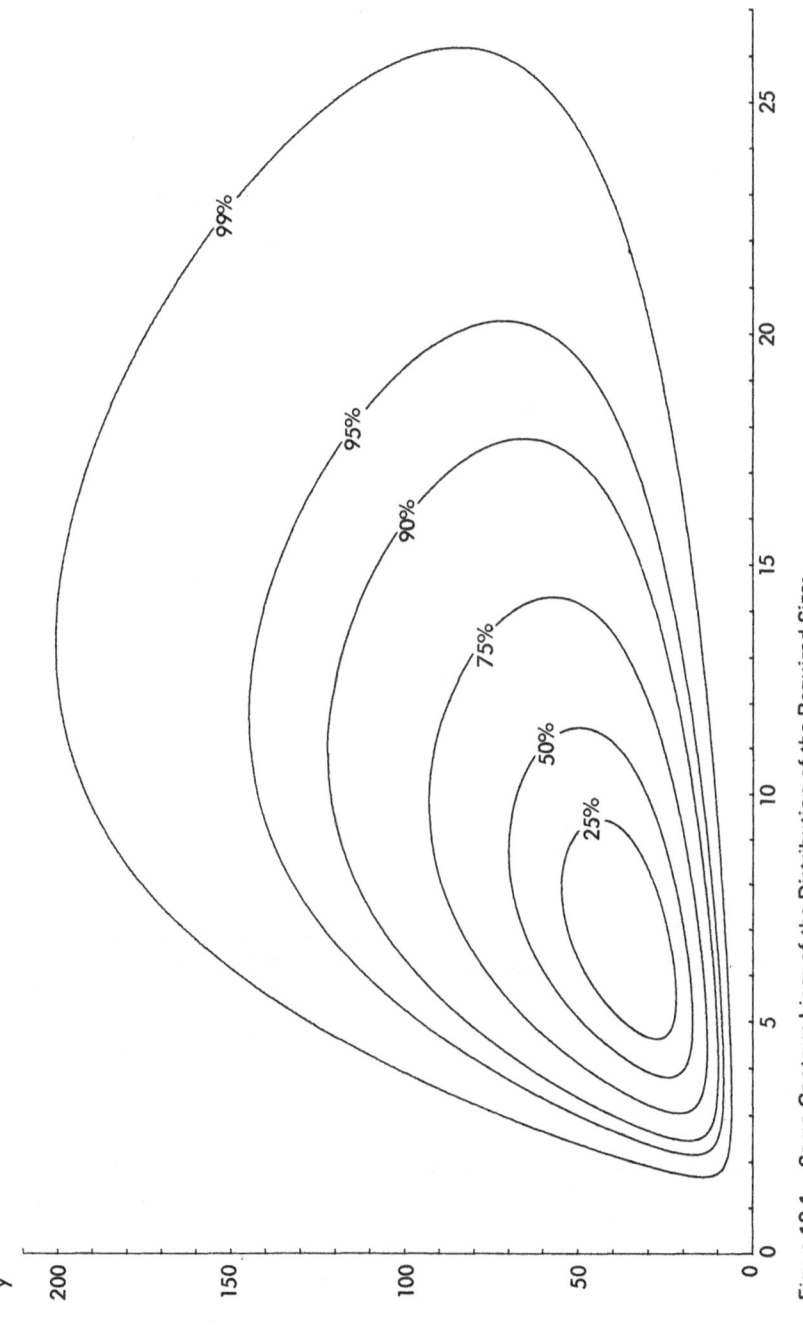

Figure 18.1.   Some Contour Lines of the Distribution of the Required Sizes

Table 18.1. Some Optimal Size Patterns Based on the Grid Restriction

$n = m$	i	x_i	y_i	Probabilities Service Areas								Coverage	$E(L)$
1	1	10.92	63.01	.55								.552	670
2	1	9.02	51.67	.14	.14							.785	456
	2	13.96	85.99	.37	.13								
3	1	8.06	45.43	.04	.07	.05						.874	343
	2	11.38	69.12	.12	.12	.06							
	3	15.96	102.02	.27	.12	.04							
4	1	7.44	41.29	.01	.03	.03	.03					.917	275
	2	10.04	59.91	.04	.06	.05	.03						
	3	13.06	82.21	.10	.09	.06	.03						
	4	17.48	114.56	.20	.10	.04	.01						
6	1	6.63	35.95	.00	.01	.01	.01	.01	.01			.956	196
	2	8.56	49.55	.01	.02	.02	.02	.02	.01				
	3	10.47	63.69	.02	.03	.03	.03	.02	.01				
	4	12.67	80.32	.04	.05	.04	.03	.02	.01				
	5	15.49	101.93	.07	.06	.05	.03	.02	.01				
	6	19.75	133.86	.13	.07	.04	.02	.01	.00				
8	1	6.12	32.56	.00	.00	.00	.01	.01	.01	.00	.00	.973	152
	2	7.71	43.60	.00	.00	.01	.01	.01	.01	.01	.00		
	3	9.18	54.31	.00	.01	.01	.01	.01	.01	.01	.01		
	4	10.71	65.85	.01	.02	.02	.02	.02	.01	.01	.00		
	5	12.44	79.13	.02	.03	.03	.02	.02	.01	.01	.00		
	6	14.52	95.39	.03	.04	.03	.03	.02	.01	.01	.00		
	7	17.27	116.89	.05	.05	.04	.03	.02	.01	.00	.00		
	8	21.44	148.66	.09	.05	.03	.02	.01	.00	.00	.00		

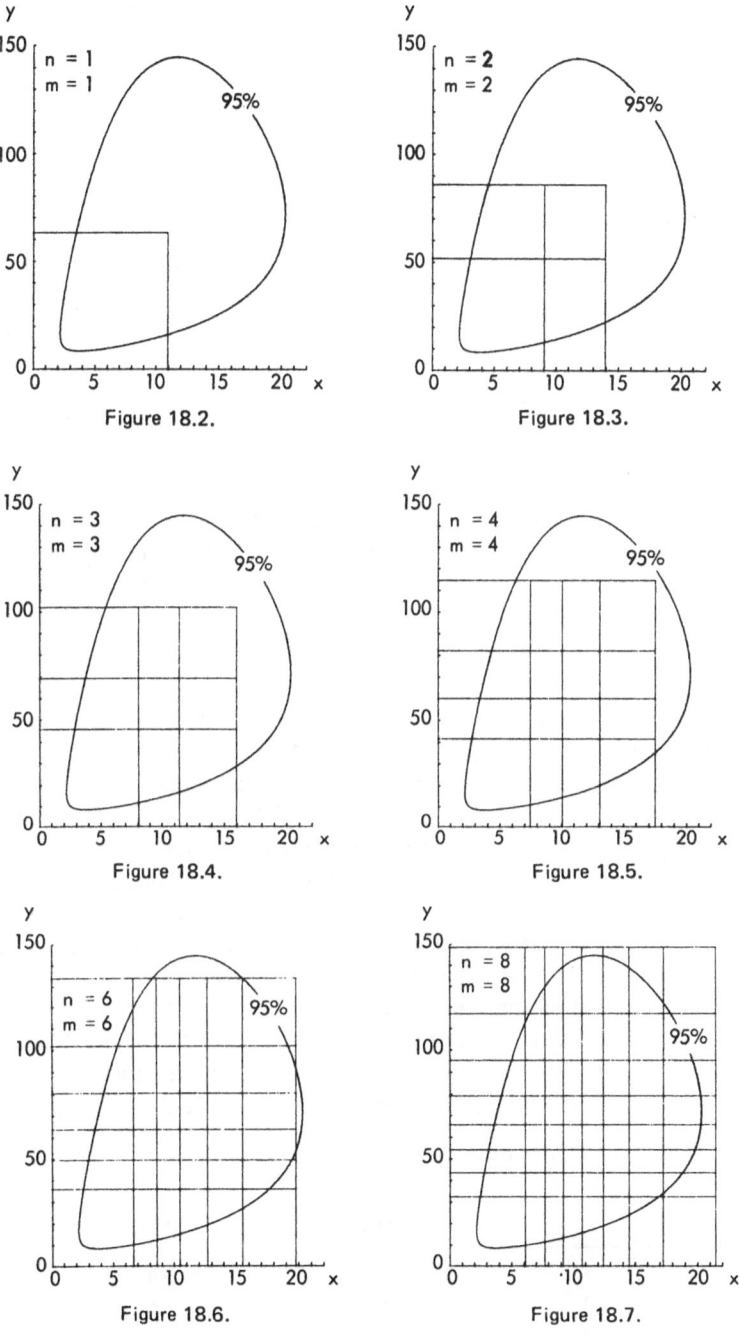

Figure 18.2.

Figure 18.3.

Figure 18.4.

Figure 18.5.

Figure 18.6.

Figure 18.7.

Figures 18.2-18.7. The Graphical Representation of the Size Patterns Listed in Table 18.1

193

Studying the functional relationship between the chosen number of sizes and the expected adaptation loss we found that this relationship—for larger values of (nm)—can be approximated surprisingly well by a function of the form

$$E(L) \cong a(nm)^{\gamma} \quad \gamma < 0. \tag{18.4}$$

To estimate a and γ the expected adaptation losses of optimal size patterns were calculated for a large number of values of n and m. For combinations of n and m that led to the same value of the product nm, we took the combination with the smallest expected adaptation loss. Using least squares regression we found the following result

$$E(L) \cong 882 \, (nm)^{-.42} \quad 9 \leqslant nm \leqslant 64. \tag{18.5}$$

The correlation coefficient between E(L) and (nm) equals .99 indicating a good fit of the function. The function (18.5) can be used—in combination with the function that relates the production costs to the number of sizes—to estimate the optimal value of (nm), say $(nm)^{opt}$. Thereafter, the optimal values of n and m can be found by calculating the expected adaptation loss for every combination of n and m whose product is in the neighborhood of $(nm)^{opt}$.

In Sec. 18.4 an example of the determination of the optimal number of sizes for flat steel sections will be given. However, for this purpose we will employ the generalized grid restriction instead of the grid restriction. The reason is that the efficiency of the grid restriction decreases rather fast if the number of sizes that has to be chosen increases. For small values of the product (nm), the probabilities of the service areas are all of a reasonable magnitude. For (nm) larger than 9, however, we see that the probabilities of some service areas become relatively very small. The probability of the service area of size (x_1, y_6) if $n = m = 6$, for instance equals .002; deleting this size from the size pattern increases the expected adaptation loss with .5 or .3% only.

To obtain a better solution for this specific case (i.e. 36 sizes have to be chosen) the generalized grid restriction can be applied. The values of the parameters n and $m_{k,i}$ ($k = 1, 2, i = 1, \ldots, n$) were determined using the procedure described in Sec. 16.3 and they are inserted into Table 18.2.

Table 18.2. The Values of $m_{1,i}$ and $m_{2,i}$ ($i = 1, \ldots, 6$) of the Generalized Grid

i	$m_{1,i}$	$m_{2,i}$
1	1	4
2	1	6
3	1	7
4	1	7
5	2	7
6	2	7

The optimal values of the size coordinates and the probabilities of the service areas are listed in Table 18.3. The size pattern is plotted in Fig. 18.8. Roughly speaking, the benefits of the use of the generalized grid restriction are to be found in the replacing of some sizes from the upper left and lower right corner of the grid to some other location. As can be seen when comparing the size patterns displayed in Figs. 18.6 and 18.8 with each other, six sizes have been replaced. As a consequence, the expected adaptation loss has decreased from 196 to 189 or by 3.7%. Another way to compare the performances of the two restrictions is to compute how many sizes are needed to realize an expected adaptation loss of 196 when employing the generalized grid restriction. Some calculations showed that a size pattern consisting of 33 sizes has an expected adaptation loss slightly less than this amount. Expressed as a percentage of the number of sizes that were needed under the grid restriction to realize an expected adaptation loss of 196, the savings due to the use of the generalized grid restriction are therefore (3/36) 100% = 8.3%.

Having considered the restricted selection of the sizes, we now turn to the free selection of the sizes. Employing the model discussed in Sec. 17.4, a number of size patterns satisfying the recursion formulae have been computed for n (the number of sizes) equal to respectively 2, 3, 4, 5, 8 and 20. No size patterns were calculated for larger values of n because of the fact that the differences between the expected adaptation losses of size patterns based upon the generalized grid restriction and those based upon unrestricted selection of the sizes then become relatively very small. In practice this will certainly lead to the use of the generalized grid restriction for values of n beyond 20, because size patterns chosen according to this restriction are easier to use and will usually have lower production and distribution costs.

To generate the size patterns we used starting sets selected randomly, except for the case n = 20. The solutions with the smallest expected adaptation losses

Table 18.3. The Optimal Size Pattern for the Grid Structure Listed in Table 18.2

i	x_i	y_i	Probabilities Service Areas						Coverage	E(L)
1	6.36	33.49	–	–	.01	.01	.01	.01		
2	8.23	45.73	–	.01	.01	.02	.01	.01		
3	10.21	58.62	–	.03	.02	.02	.02	.01	.962	189
4	12.56	73.15	.02	.03	.04	.03	.02	.01		
5	15.42	89.66	.03	.05	.05	.03	.02	.01		
6	19.71	111.60	.06	.06	.05	.03	.02	.01		
7	–	141.32	.10	.06	.03	.02	–	–		

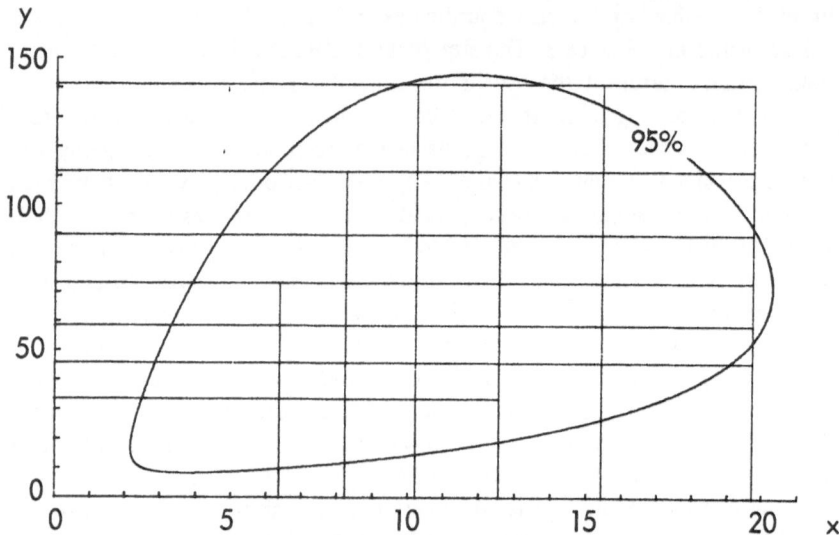

Figure 18.8.   The Graphical Representation of the Size Pattern Listed in Table 18.3

are listed in Table 18.4 and plotted in Figs. 18.9-18.13 and 18.17. For the specific case n = 8 we found in total 11 different size patterns satisfying the recursion formulae. The size pattern displayed in Fig. 18.14 has the largest expected adaptation loss among these, i.e. 354. The difference between this figure and the minimum expected adaptation loss found (i.e. 337) is therefore only about 5%. Although, of course, it is not certain that the global minimum equals the local minimum of 337, the relatively small difference between the expected adaptation losses of the "best" and the "worst" size pattern indicates that the global minimum most probably will be in the very close neighborhood of 337. This opinion is also based on the structure of the size pattern displayed in Fig. 18.13. Apart from size (x_1,y_1) and size (x_2,y_2), the sizes have a specific relation to each other which is also illustrated by Fig. 18.15. Size patterns possessing such a structure in general appear to have a smaller expected adaptation loss than other size patterns. This is not so only when using the log normal distribution with the quadratic loss function. If we look for instance at Fig. 17.8 we see that the size pattern with the smallest expected adaptation loss if the probability density function of the required sizes is the bivariate normal distribution and the loss function is linear, has nearly the same structure. Of course there are exceptions in which a structure of this type is certainly not optimal, for instance if ρ approaches unity, but in most cases it gives very good results.

Table 18.4. Some Optimal Size Patterns Based on Free Selection of the Sizes

n	i	x_i	y_i	p_i	Structure Size Pattern								E(L)
2	1	9.68	53.31	.42	2	2							
	2	12.67	77.22	.29	1	2							544
3	1	8.91	47.76	.33	3	3	3						
	2	11.37	66.00	.26	2	2	3						478
	3	13.69	86.09	.19	1	2	3						
4	1	8.38	44.01	.27	4	4	4	4					
	2	10.48	59.26	.23	3	3	3	4					437
	3	12.45	74.44	.18	2	2	3	4					
	4	14.37	92.45	.13	1	2	3	4					
5	1	8.33	43.66	.26	5	5	5	5	0				
	2	11.28	54.05	.22	4	4	5	5	0				
	3	15.19	67.07	.12	4	4	3	3	3				403
	4	9.73	81.88	.14	2	2	2	3	3				
	5	13.31	100.96	.10	1	2	2	3	3				
8	1	7.49	37.84	.18	8	8	8	8	8	8	0	0	
	2	9.08	48.93	.17	7	7	7	7	8	8	0	0	
	3	11.51	53.59	.14	7	7	7	7	6	6	6	6	
	4	14.11	63.59	.09	5	5	5	7	6	6	6	6	337
	5	9.12	76.37	.11	5	5	5	4	4	4	4	6	
	6	16.78	78.27	.06	3	3	3	3	3	4	4	6	
	7	11.09	96.33	.08	2	2	3	3	3	4	4	6	
	8	14.08	112.84	.05	1	2	3	3	3	4	4	6	

n	i	x_i	y_i	p_i	i	x_i	y_i	p_i	E(L)
20	1	6.56	31.71	.097	11	19.44	74.49	.022	
	2	9.32	35.96	.076	12	12.61	75.79	.044	
	3	11.83	39.69	.059	13	8.07	79.11	.041	
	4	6.86	44.55	.091	14	14.15	87.87	.034	
	5	14.25	49.37	.047	15	9.39	90.61	.040	233
	6	10.33	50.51	.060	16	16.38	100.30	.022	
	7	7.61	56.29	.082	17	10.49	105.42	.032	
	8	16.74	61.53	.032	18	18.61	116.01	.016	
	9	9.23	62.57	.057	19	11.81	122.26	.024	
	10	11.43	66.13	.053	20	13.76	137.88	.018	

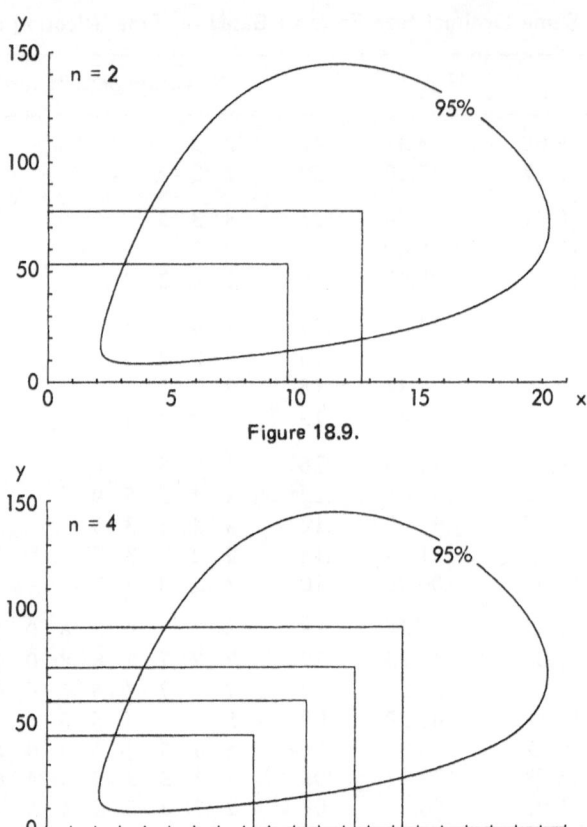

Figure 18.9.

Figure 18.11.

Figures 18.9–18.13. The Graphical Representation of the Size
Patterns Listed in Table 18.4

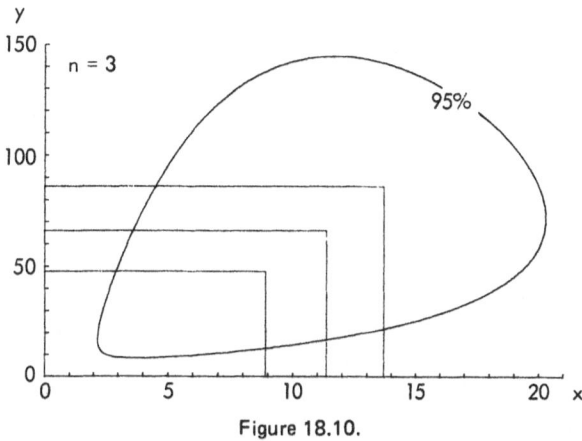

Figure 18.10.

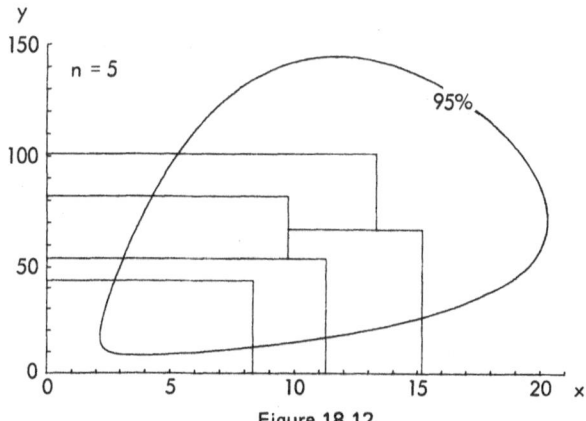

Figure 18.12.

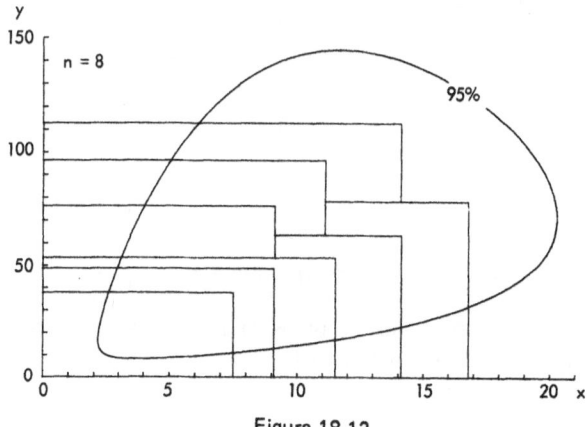

Figure 18.13.

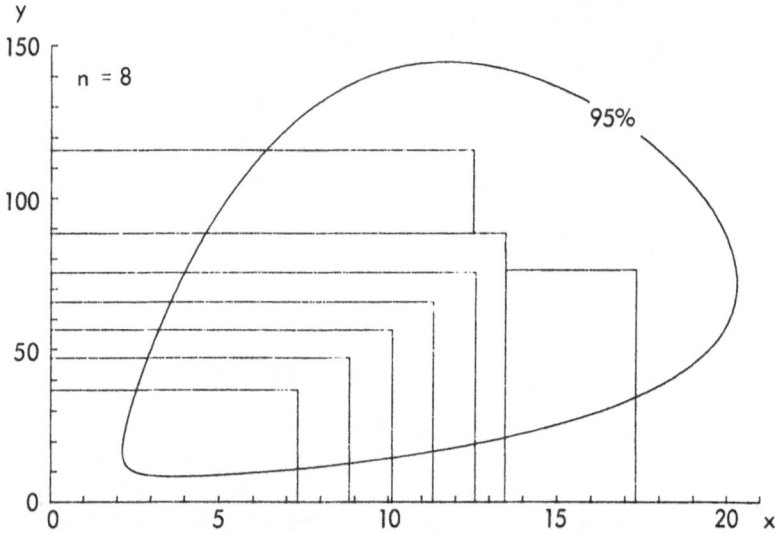

Figure 18.14. The Size Pattern with the Largest Expected Adaptation Loss among the Size Patterns Consisting of 8 Sizes

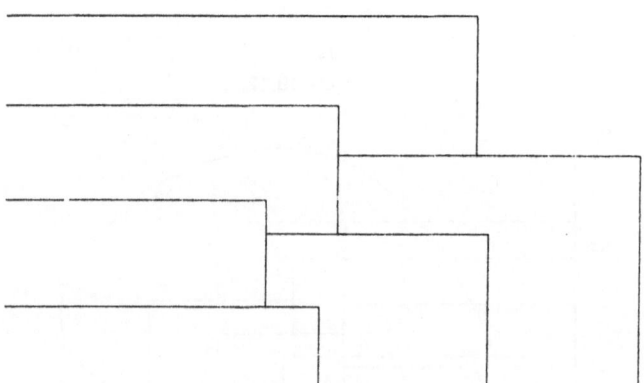

Figure 18.15. The Structure Which Usually Leads to a Relatively Small Value of the Expected Adaptation Loss

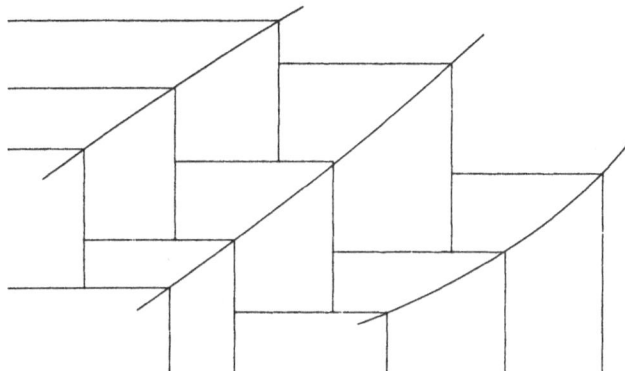

Figure 18.16. The Structure Used to Generate the Size Pattern
Displayed in Fig. 18.17

We used a slight modification of the structure displayed in Fig. 18.15 to choose a number of starting sets in order to find some (sub)optimal patterns for the case n = 20. The modified structure is displayed in Fig. 18.16. The size pattern with the smallest expected adaptation loss is displayed in Fig. 18.17. The numbers inside the service areas denote the probabilities of the service areas; as can be seen these probabilities are decreasing along each of the three lines which can be drawn to connect the sizes (see Fig. 18.16). Furthermore, it is maybe interesting to note that the expected adaptation loss of the starting set – which was chosen rather carefully – was about 20% larger than the expected adaptation loss of the size pattern displayed in Fig. 18.17. This again indicates that the use of intuitive methods to select a size pattern are to be avoided if sufficient data are available to specify a model.

Up to this point we have illustrated the various models which can be used to select a size pattern for flat steel sections. Though this is interesting from a theoretical point of view, we like to emphasize their practical importance too. For this purpose, we will in the next section compare the expected adaptation losses of some size pattern computed with one of our models with that of an existing size pattern. For the latter we have taken the size pattern of Euronorm 58–76 and it shall be shown that – under our assumptions – considerable savings can be realized by using an optimized size pattern based upon the generalized grid restriction instead of the size pattern of the Euronorm.

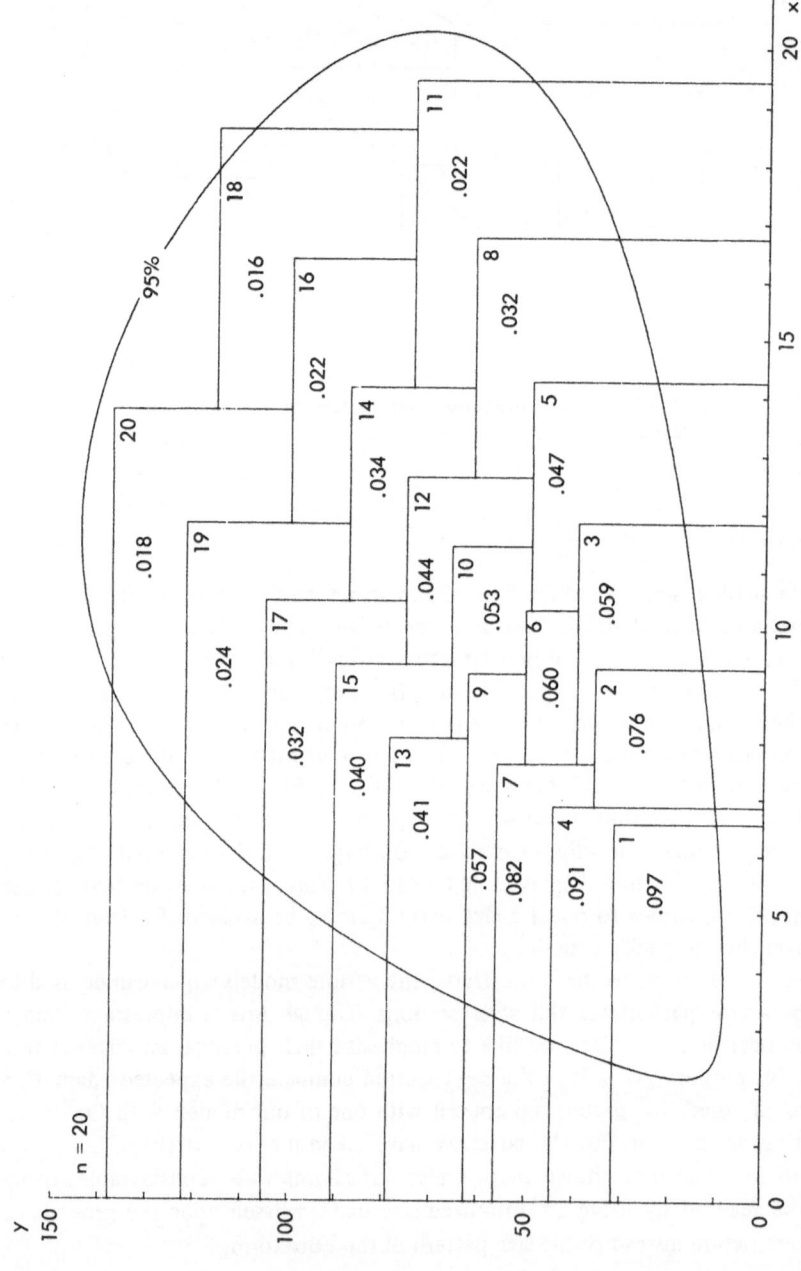

Figure 18.17. The Size Pattern with the Smallest Expected Adaptation Loss among the Size Patterns Consisting of 20 Sizes

18.3. A COMPARISON OF THE SIZE PATTERN OF EURONORM
58-76 WITH SOME OPTIMAL SIZE PATTERNS

The size pattern as given by Euronorm 58-76 consists of 126 sizes. The structure of this size pattern can be described by a generalized grid with the following values of the parameters (see Table 18.5). Furthermore, the values of the size coordinates (in mm) are listed in Table 18.6.

Table 18.5. The Values of $m_{1,i}$ and $m_{2,i}$ of the Size Pattern of the Euronorm

i	$m_{1,i}$	$m_{2,i}$	i	$m_{1,i}$	$m_{2,i}$
1	1	12	7	3	14
2	1	14	8	5	14
3	1	14	9	5	14
4	1	14	10	8	14
5	1	14	11	10	14
6	1	14			

Table 18.6. The Size Coordinates of the Size Pattern of the Euronorm

i	x_i	y_i	i	x_i	y_i
1	5	20	8	25	60
2	6	25	9	30	70
3	8	30	10	40	80
4	10	35	11	50	90
5	12	40	12	–	100
6	15	45	13	–	120
7	20	50	14	–	150

The size pattern is displayed in Fig. 18.8 together with the 99% contour line of the distribution of the required sizes. We see from this figure that a number of sizes—more specifically, the sizes with an x coordinate equal to 30, 40 or 50— are located far outside the 99% contour line. This rather strange fact cannot be explained by a shift of the demand from 1962—the year in which the data were gathered on basis of which the parameters of the density function were estimated—to 1976. The previous Euronorm (Euronorm 58-64) namely also contained sizes with the same x coordinates.

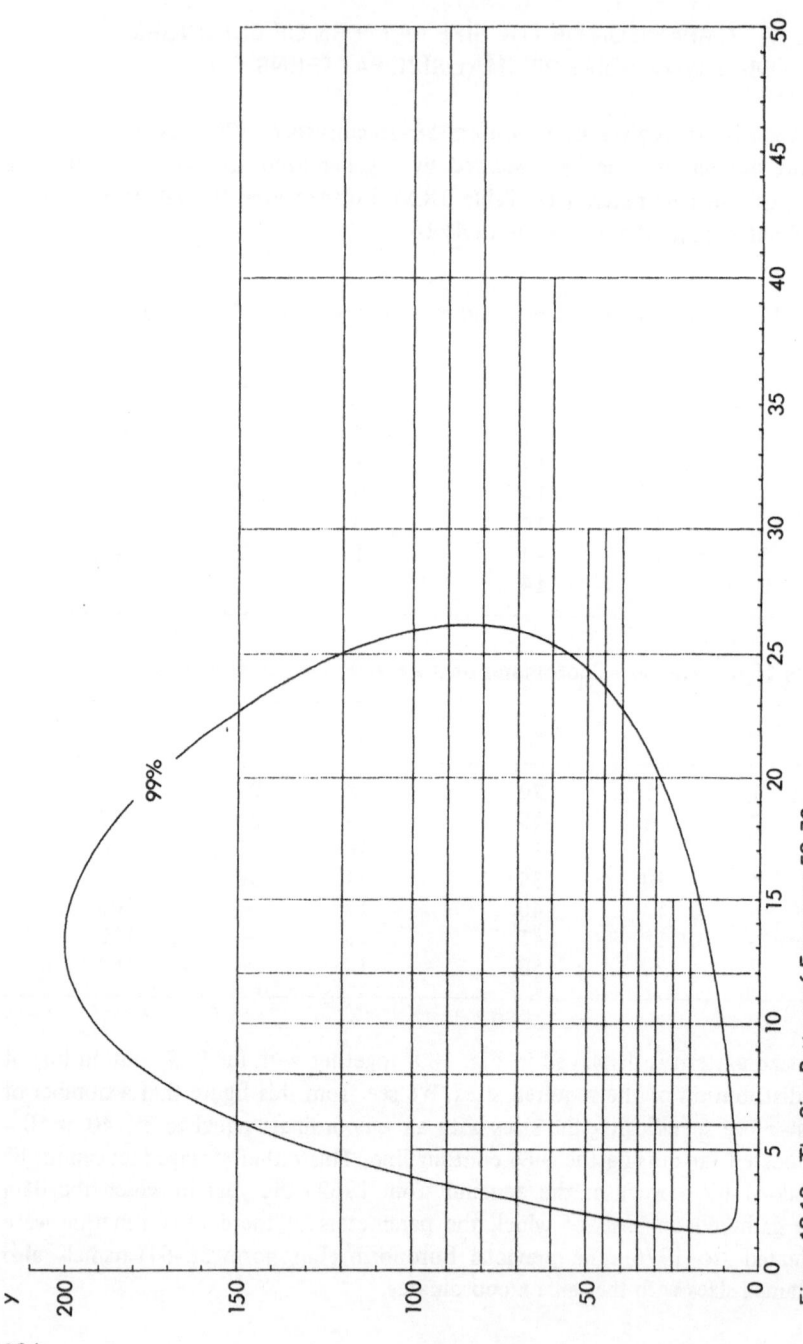

Figure 18.18.   The Size Pattern of Euronorm 58–76

204

A motivation for the choice of the large x coordinates might have been that the designers of the Euronorm aimed at a large coverage of the size pattern in order to be sure that nearly all consumers could be served. This however ought to imply that also y_m has a large value which is not the case; applying namely transformation (7.3) to y_m — thus transforming y to a standard normally distributed variable — we find $u_m = 2.14$. This value is rather low as compared to the value of v_n which equals 4.36. Another possible explanation is that the loss of not being served differs in the following three situations.

1. $x > x_n$ and $y < y_m$,
2. $y > y_m$ and $x < x_n$,
3. $x > x_n$ and $y > y_m$.

Only if it is assumed that the loss of not being served in situation 1 and/or 3 is many times larger than that in situation 2 is it possible to arrive at a size (x_n, y_m) equal to (50, 150). Although — in our opinion — this assumption is rather questionable, we shall accept it to make the comparison between the size patterns possible.

We shall use for this comparison some size patterns generated with the generalized grid restriction. In all the size patterns the size in the right upper corner of the generalized grid is chosen equal to (50, 150). Consequently, the derivatives with respect to x_n and y_m are not used, which implies that we do not consider the loss of not being served.

Savings — in terms of the expected adaptation loss — can be realized in the following three ways.

1. Using the same generalized grid as the Euronorm, the values of the size coordinates can be optimized.
2. Assuming 126 sizes have to be chosen, it is possible to reduce the expected adaptation loss by the choice of a better generalized grid.
3. It is also possible to reduce the production costs by finding a size pattern with fewer sizes than that of the Euronorm but with the same expected adaptation loss.

Attention shall be paid to points 1 and 3. Starting with point 1, we have first calculated the coverage and the (conditional) expected adaptation loss[2] associated with the size pattern described in the Euronorm. These were equal to respectively .984 and 123.

---

[2] The expected adaptation losses in the rest of this section are computed under the condition that the consumer is served.

Next, the optimal size pattern has been calculated employing the same generalized grid as that of the Euronorm and with size (x_{11}, y_{14}) fixed at (50, 150). This size pattern is inserted into Table 18.7 and displayed in Fig. 18.19. From the table we see that the expected adaptation loss equals 100, which means – as compared to the size pattern of the Euronorm – a reduction of 23 or 19%. Furthermore, since a running mm of flat steel contains on the average

$$\left(\int_0^{50} \int_0^{150} xy\, h(x,y)\, dy\, dx \right) / .984 = 557 \text{ mm}^3$$

steel, we may also say that an optimal choice of the sizes leads to a reduction of (23/557) 100% is about 4% in the usage of steel.

A question that has been neglected so far is how the expected adaptation loss is influenced if the sizes are rounded off to whole numbers as is done with the size pattern of the Euronorm. Such a size pattern is listed in Table 18.8. The expected adaptation loss of this size pattern equals 104, implying an increase of 4 or 4%. Experience with various other size patterns also teaches that rounding off the sizes usually increases the expected adaptation loss by a relatively small amount only. This, however, implies nothing with respect to the total economic loss – caused by the rounding off process – which might be considerable. We are therefore of the opinion that automatically rounding off the sizes is not to be recommended. Instead one ought to perform a sensitivity analysis to determine the number of significant digits of the sizes. Next it must be decided whether it is justified to decrease the number of significant digits by rounding off, taking into consideration the total economic loss incurred by the rounding off process. If this loss is too large it might be better to design a code system for the sizes, as for instance is done in the textile industry.

Table 18.7. The Optimal Size Pattern with the Same Generalized Grid as the Euronorm

		E(L) = 100,	Coverage = .984		
i	x_i	y_i	i	x_i	y_i
1	5.97	24.71	8	18.28	69.12
2	7.46	31.32	9	22.25	77.53
3	8.83	37.08	10	29.04	87.04
4	10.22	42.80	11	50.00	98.24
5	11.75	48.51	12	–	111.68
6	13.52	54.97	13	–	128.18
7	15.64	61.78	14	–	150.00

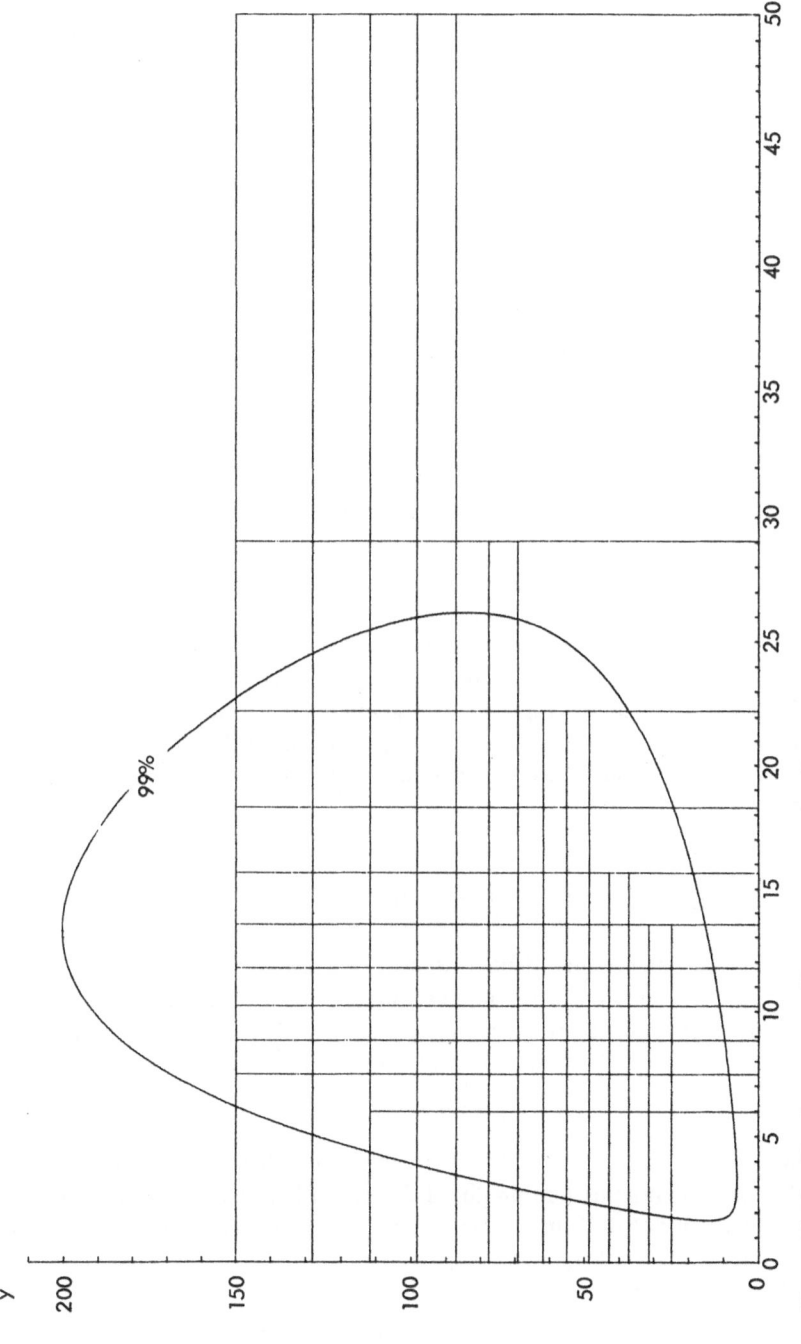

Figure 18.19. The Optimal Size Pattern with the Same Generalized Grid as the Size Pattern of the Euronorm

207

Table 18.8. The Size Pattern of Table 18.7 but Rounded Off to Nearest Integers

		E(L) = 104,		Coverage = $.984$	
i	x_i	y_i	i	x_i	y_i
1	6	25	8	18	69
2	7	31	9	22	78
3	9	37	10	29	87
4	10	43	11	50	98
5	12	49	12	-	112
6	14	55	13	-	128
7	16	62	14	-	150

Next, we discuss how savings can be realized by selecting a size pattern with fewer sizes than that of the Euronorm, under the constraint that it must have the same expected adaptation loss. To obtain a size pattern satisfying this restriction, the procedure described in Sec. 16.3 has been applied to two size patterns chosen with the grid restriction. The first size pattern consisted of 121 sizes, with n = m = 11, whereas the second size pattern consisted of 120 sizes, with n = 12 and m = 10. The latter choice appeared to be the best; the end result is the size pattern listed in Table 18.9. A graphical representation of this size pattern is given in Fig. 18.20. As can be seen from the table, the size pattern consists of 78 sizes, which means a reduction of 48 sizes or 38% as compared to the size pattern of the Euronorm.

18.4   THE OPTIMAL NUMBER OF SIZES

To determine the optimal number of sizes, we have to express the total production costs as a function of the number of sizes (n). Under the assumption that the total production costs vary proportionally with the quantity of flat steel produced, we may write these costs — at a production of 126 sizes — as,

$$d(126) = k(557 + E(L)) \qquad (18.6)$$

where k equals the product of the manufacturing costs per mm^3 of flat steel and the number of running mm produced. The term between brackets denotes the average quantity of steel (in mm^3) per running mm. To incorporate the influence the number of sizes has on the total production costs, we multiply (18.6) by $(n/126)^\epsilon$, which gives

$$d(n) = k(557 + E(L))(n/126)^\epsilon. \qquad (18.7)$$

Employing the generalized grid restriction, the expected adaptation losses of a number of optimal size patterns—with the size in the right upper corner fixed at (50, 150)—have been calculated to estimate the relationship between E(L) and n. We obtained, using least squares regression,

$$E(L) = 1972 \, n^{-.64} \qquad 36 \leqslant n. \qquad (18.8)$$

According to Sittig (1963), the value of ϵ is approximately equal to .125 for the production of flat steel. The production costs for ϵ = .125 are listed in Table 18.10 as a function of n. To facilitate comparisons, the production costs at a production of 126 sizes are put equal to 100. As can be seen from the table, the savings due to the optimal choice of n (i.e. n = 66) are approximately 1%. Since the curve is rather flat in the neighborhood of the optimal n, these savings hardly change if another value of n is chosen between 50 and 90.

Table 18.9. The Optimal Size Pattern with 78 Sizes (n = 12, m = 10)

			E(L) = *123*,		*Coverage = .984*			
i	$m_{1,i}$	$m_{2,i}$	x_i	y_i	i	$m_{1,i}$	$m_{2,i}$	x_i	y_i
1	1	3	5.26	27.39	7	3	10	14.80	87.08
2	1	7	6.60	35.95	8	3	10	17.39	102.33
3	1	9	8.04	44.67	9	4	10	21.01	122.12
4	1	10	9.53	53.45	10	6	10	26.46	150.00
5	2	10	11.07	63.12	11	9	10	33.46	–
6	2	10	12.82	73.93	12	10	10	50.00	–

Table 18.10. The Production Costs as a Function of n

n	*Production Costs*	n	*Production Costs*
126	100	70	98.78
110	99.55	66	98.77
100	99.29	60	98.79
90	99.06	50	99.02
80	98.88	40	99.61

210

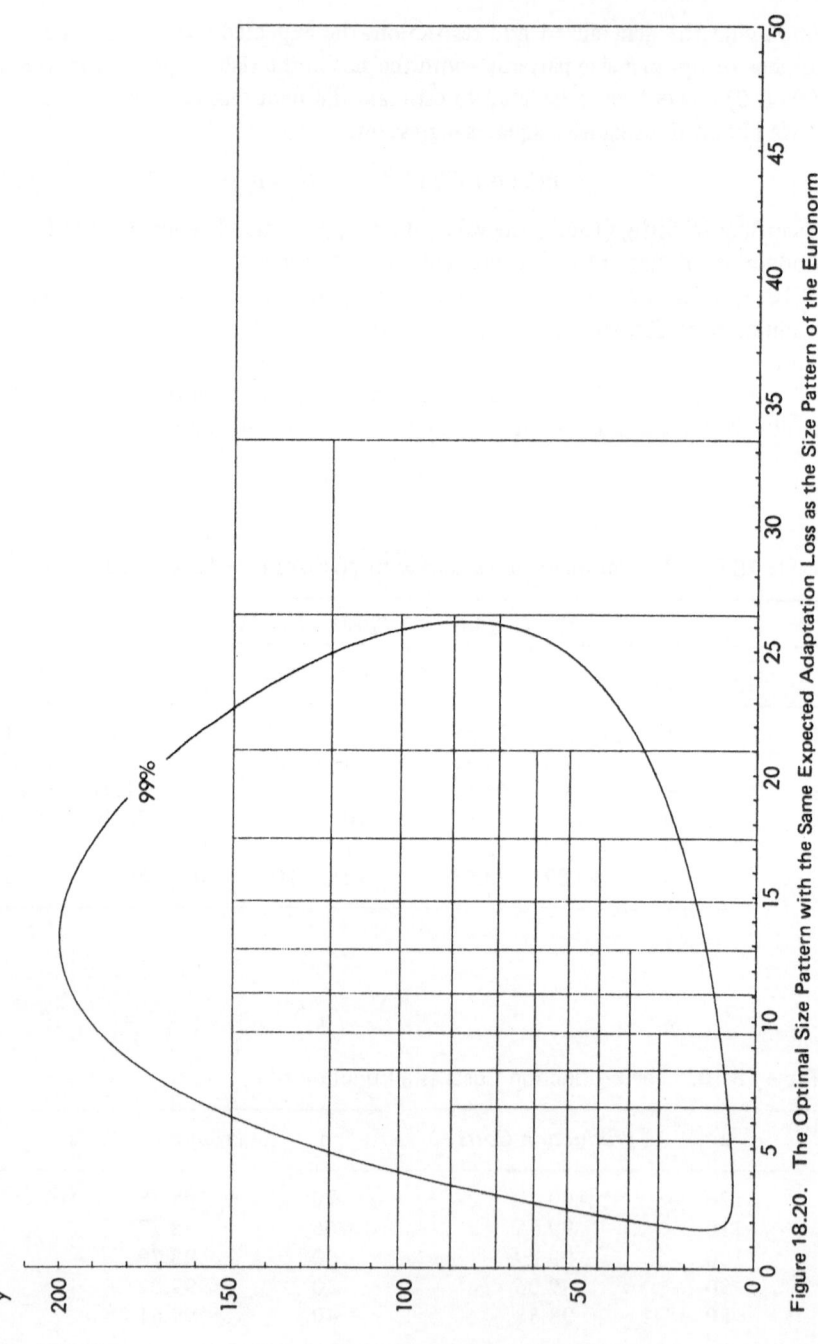

99%

y

200

150

100

50

0

0   5   10   15   20   25   30   35   40   45   50   x

Figure 18.20. The Optimal Size Pattern with the Same Expected Adaptation Loss as the Size Pattern of the Euronorm

18.5 CONCLUSIONS

Table 18.11 contains the results of the various models that have been considered in the previous two sections. The total production costs and the expected adaptation loss of the size pattern of the Euronorm are both put equal to 100.

From the table it will be clear, that—under our assumptions—a considerable reduction of the total production costs can be realized (i.e. 6%) when standardizing flat steel using the discussed model. In short, these assumptions were

1. The consumer applies shift rule O with respect to the width and thickness when buying flat steel.
2. A consumer who needs more than one size does not try to reduce his adaptation loss by splitting up a standardized size into two or more required sizes.
3. The loss function is of the form (18.3).
4. The probability density function of the required sizes is bivariate log normal with parameter values equal to the estimated parameter values.
5. The standardization process of flat steel does not depend on other standardization processes.
6. The value of ϵ equals .125 for the production of flat steel.

When adopting one of the models we propose, it is, of course, essential to check these assumptions critically. With respect to point 5, for instance, it is very well possible that one or more specific sizes of the Euronorm were chosen under the restriction that they must connect precisely to other (already standardized) products. To take such restrictions into account a slight modification of the model is necessary. The derivatives with respect to the size coordinates

Table 18.11. Summary of Results

No. of Sizes	Size Pattern Chosen With	Total Estimated Production Costs	Expected Adaptation Loss
126	Euronorm	100	100
126	Optimized grid of Euronorm	96.6	81.3
78	Optimal grid	94.0	100
66	Optimal grid, Optimal n	93.9	109.8

which are fixed in advance of the optimization process namely have to be omitted and the values of these sizes coordinates must be substituted in the remaining recursion formulae.

Finally, we remark that although the models that have been introduced in this and the previous chapters are by no means perfect and complete we are convinced that they can be of great use when standardizing certain products. Intensive research directed towards the development of more elaborate models is of importance, as in our opinion the building of a theoretical framework for the problem of standardization will lead to considerable economic benefits.

APPENDICES

A TABLES OF OPTIMAL SIZE PATTERNS

Table A.1. Some Optimal Transformed Size Patterns, Computed with the Normal Distribution, the Linear Loss Function and Shift Rule O (See Sec. 6.3)

n,E(L), Cov.	i				Sizes			
		$B/k\sigma = .5$	$B/k\sigma = 1$	$B/k\sigma = 1.5$	$B/k\sigma = 3$	$B/k\sigma = 5$	$B/k\sigma = 10$	$B/k\sigma = 25$
n = 1	1	-1.57	-.30	.21	.88	1.26	1.69	2.15
E(L)/B		.99	.88	.76	.52	.37	.22	.10
Cov.		.06	.38	.58	.81	.90	.95	.98
n = 2	1	-1.12	-.43	-.17	.12	.27	.42	.57
	2	-.51	.49	.92	1.50	1.85	2.24	2.67
E(L)/B		.92	.70	.54	.32	.21	.12	.05
Cov.		.31	.69	.82	.93	.97	.99	1.00
n = 3	1	-1.08	-.64	-.49	-.32	-.23	-.15	-.07
	2	-.46	.16	.39	.67	.82	.96	1.11
	3	.06	.92	1.31	1.84	2.16	2.53	2.94
E(L)/B		.83	.56	.42	.24	.15	.08	.03
Cov.		.52	.82	.91	.97	.98	.99	1.00
n = 4	1	-1.14	-.84	-.73	-.61	-.56	-.50	-.45
	2	-.53	-.12	.03	.20	.29	.37	.45
	3	-.04	.52	.73	1.00	1.14	1.28	1.42
	4	.43	1.21	1.57	2.07	2.37	2.73	3.12
E(L)/B		.73	.46	.34	.18	.12	.06	.03
Cov.		.67	.89	.94	.98	.99	1.00	1.00

n = 5							
1	-.71	-.75	-.79	-.83	-.92	-.99	-1.22
2	.05	.00	-.05	-.11	-.23	-.34	-.64
3	.77	.69	.61	.52	.36	.21	-.17
4	1.63	1.50	1.36	1.23	.98	.77	.26
5	3.25	2.87	2.53	2.24	1.77	1.43	.70
E(L)/B	.02	.05	.09	.15	.28	.39	.65
Cov.	1.00	1.00	.99	.99	.96	.92	.76

n = 6							
1	-.91	-.94	-.97	-1.00	-1.07	-1.13	-1.31
2	-.22	-.26	-.30	-.35	-.44	-.52	-.75
3	.37	.32	.26	.20	.09	-.02	-.31
4	.99	.92	.84	.75	.59	.46	.09
5	1.80	1.66	1.53	1.40	1.16	.97	.49
6	3.35	2.98	2.65	2.37	1.92	1.60	.91
E(L)/B	.02	.04	.08	.13	.24	.34	.58
Cov.	1.00	1.00	1.00	.99	.97	.94	.82

n = 7							
1	-1.07	-1.09	-1.11	-1.14	-1.19	-1.24	-1.39
2	-.43	-.46	-.50	-.53	-.60	-.66	-.85
3	.09	.05	.01	-.04	-.13	-.21	-.43
4	.60	.55	.49	.43	.32	.22	-.06
5	1.17	1.09	1.01	.93	.78	.65	.30
6	1.93	1.80	1.67	1.54	1.31	1.13	.67
7	3.44	3.08	2.76	2.48	2.04	1.73	1.08
E(L)/B	.01	.03	.07	.11	.21	.30	.53
Cov.	1.00	1.00	1.00	.99	.98	.96	.86

Table A.2. Some Optimal Transformed Size Patterns, Computed with the Normal Distribution, the Quadratic Loss Function ($k_1 = k_2$) and Shift Rule T (See Sec. 6.4)

The sizes listed are: $v_{(n+1)/2} - v_n$ for odd values of n,
$v_{1+n/2} - v_n$ for even values of n

n	E(L)						Sizes							
		1	2	3	4	5	6	7	8	9	10	11	12	13
1	1.000	0.000												
2	0.363	0.798												
3	0.190	0.000	1.224											
4	0.117	0.453	1.510											
5	0.080	0.000	0.765	1.724										
6	0.058	0.318	1.000	1.894										
7	0.044	0.000	0.561	1.188	2.033									
8	0.035	0.245	0.756	1.344	2.152									
9	0.028	0.000	0.444	0.919	1.476	2.255								
10	0.023	0.200	0.610	1.058	1.591	2.345								
11	0.019	0.000	0.367	0.752	1.179	1.693	2.426							
12	0.016	0.168	0.512	0.877	1.286	1.783	2.498							
13	0.014	0.000	0.314	0.638	0.987	1.381	1.864	2.564						
14	0.012	0.146	0.441	0.750	1.086	1.467	1.939	2.625						
15	0.011	0.000	0.274	0.555	0.851	1.175	1.546	2.006	2.681					
16	0.010	0.128	0.388	0.657	0.942	1.256	1.618	2.069	2.733					
17	0.008	0.000	0.243	0.491	0.749	1.026	1.331	1.684	2.127	2.781				
18	0.008	0.115	0.346	0.584	0.834	1.102	1.400	1.746	2.181	2.826				
19	0.007	0.000	0.218	0.440	0.670	0.911	1.173	1.464	1.803	2.231	2.868			
20	0.006	0.104	0.313	0.526	0.748	0.984	1.238	1.523	1.857	2.279	2.908			
21	0.006	0.000	0.198	0.399	0.606	0.821	1.050	1.300	1.579	1.907	2.323	2.945		
22	0.005	0.095	0.285	0.479	0.679	0.889	1.113	1.357	1.631	1.955	2.365	2.981		
23	0.005	0.000	0.182	0.365	0.553	0.748	0.953	1.171	1.411	1.681	1.999	2.405	3.015	
24	0.004	0.087	0.262	0.440	0.622	0.812	1.012	1.226	1.462	1.728	2.042	2.443	3.047	
25	0.004	0.000	0.167	0.337	0.509	0.687	0.872	1.068	1.278	1.510	1.772	2.082	2.479	3.078

Table A.3. Some Optimal Transformed Size Patterns, Computed with the Normal Distribution, the Linear Loss Function ($k_1 = k_2$) and Shift Rule T (See Sec. 6.5)

The sizes listed are: $v_{(n+1)/2} - v_n$ for odd values of n, $v_{1+n/2} - v_n$ for even values of n

n	E(L)	1	2	3	4	5	6	7	8	9	10	11	12	13
1	0.798													
2	0.473	0.674												
3	0.340	0.000	1.029											
4	0.266	0.378	1.266											
5	0.218	0.000	0.636	1.442										
6	0.186	0.264	0.831	1.581										
7	0.161	0.000	0.456	0.986	1.695									
8	0.143	0.203	0.626	1.114	1.792									
9	0.128	0.000	0.367	0.760	1.223	1.876								
10	0.116	0.166	0.504	0.874	1.317	1.950								
11	0.106	0.000	0.304	0.622	0.974	1.400	2.016							
12	0.098	0.140	0.423	0.724	1.061	1.474	2.075							
13	0.091	0.000	0.260	0.527	0.814	1.140	1.541	2.129						
14	0.085	0.121	0.365	0.620	0.895	1.210	1.601	2.178						
15	0.079	0.000	0.227	0.459	0.702	0.969	1.275	1.657	2.224					
16	0.074	0.107	0.321	0.542	0.777	1.035	1.334	1.708	2.266					
17	0.070	0.000	0.202	0.406	0.619	0.845	1.096	1.388	1.755	2.305				
18	0.066	0.096	0.287	0.483	0.688	0.908	1.153	1.438	1.799	2.341				
19	0.063	0.000	0.182	0.365	0.554	0.752	0.966	1.205	1.485	1.840	2.376			
20	0.060	0.087	0.260	0.436	0.618	0.811	1.020	1.254	1.529	1.879	2.408			
21	0.057	0.000	0.166	0.332	0.502	0.678	0.866	1.070	1.300	1.570	1.915	2.438		
22	0.055	0.080	0.238	0.398	0.562	0.734	0.917	1.117	1.342	1.608	1.949	2.467		
23	0.052	0.000	0.152	0.304	0.459	0.618	0.786	0.965	1.161	1.382	1.645	1.981	2.494	
24	0.050	0.074	0.219	0.366	0.516	0.671	0.835	1.010	1.203	1.421	1.680	2.013	2.521	
25	0.048	0.000	0.141	0.281	0.423	0.569	0.721	0.881	1.053	1.242	1.457	1.713	2.042	2.546

Sizes

219

B MATHEMATICAL DERIVATIONS

B.1 THE UNIVARIATE AND THE BIVARIATE NORMAL DISTRIBUTION

B.1.1 Notation

We denote the univariate normal distribution by $h(x)$, thus

$$h(x) = \frac{1}{\sigma\sqrt{2\pi}} e^{-((x-\mu)/\sigma)^2/2}. \tag{B.1}$$

Putting μ equal to zero and σ equal to one we have the standard normal distribution,

$$f(x) = \frac{1}{\sqrt{2\pi}} e^{-x^2/2}. \tag{B.2}$$

Distribution functions are denoted by capitals, so we have for instance,

$$F(x_i) = \int_{-\infty}^{x_i} f(x)dx. \tag{B.3}$$

Furthermore, we write the bivariate normal distribution as

$$w(x,y) = \frac{1}{2\pi\sigma_x\sigma_y\sqrt{1-\rho^2}} \exp\left[-\left(\left(\frac{x-\mu_x}{\sigma_x}\right)^2 - 2\rho\left(\frac{x-\mu_x}{\sigma_x}\right)\left(\frac{y-\mu_y}{\sigma_y}\right)\right.\right.$$
$$\left.\left. + \left(\frac{y-\mu_y}{\sigma_y}\right)^2\right)/(2(1-\rho^2))\right] \qquad (B.4)$$

Substituting $\mu_x = \mu_y = 0$ and $\sigma_x = \sigma_y = 1$ the standard bivariate normal distribution arises

$$g(x,y) = \frac{1}{2\pi\sqrt{1-\rho^2}} e^{-(1-\rho^2)\{x^2 - 2\rho xy + y^2\}/(2(1-\rho^2))}. \qquad (B.5)$$

B.1.2   Transformations and Substitutions

In part B.1 of this appendix the following transformations and substitutions are used:

$$v = (x - \mu_x)/\sigma_x, \qquad (B.6) \qquad\qquad u = (y - \mu_y)/\sigma_y \qquad (B.7)$$
$$v_i = (x_i - \mu_x)/\sigma_x, \qquad (B.8) \qquad\qquad u_j = (y_j - \mu_y)/\sigma_y, \qquad (B.9)$$
$$v_i^k = (v_i - \rho u_k)/\sqrt{1-\rho^2}, \qquad (B.10) \qquad u_j^k = (u_j - \rho v_k)/\sqrt{1-\rho^2}. \qquad (B.11)$$

B.1.3   Some Derivations (See Table B.1)[1]

a.
$$\int_{-\infty}^{x_i} x\, h(x)dx = \sigma \int_{-\infty}^{x_i} \frac{(x-\mu)}{\sigma} h(x)dx + \mu\, H(x_i). \qquad (B.12)$$

Applying transformation (B.6) and substitution (B.8) we can write (B.12) as

$$\sigma \int_{-\infty}^{v_i} vf(v)dv + \mu F(v_i) = \sigma \int_{-\infty}^{v_i} -df(v)$$
$$+ \mu\, F(v_i) = \mu F(v_i) - \sigma f(v_i). \qquad (B.13)$$

b.
$$\int_{-\infty}^{x_i} x^2 h(x)dx = \sigma^2 \int_{-\infty}^{x_i}\left(\frac{x-\mu}{\sigma}\right)^2 h(x)dx$$
$$+ 2\mu\int_{-\infty}^{x_i} x\, h(x)dx - \mu^2 \int_{-\infty}^{x_i} h(x)dx. \qquad (B.14)$$

[1] μ equals μ_x and σ equals σ_x.

Table B.1. The Formulae Derived in Appendix B

Normal Distribution	Bivariate Normal Distribution	Log Normal Distribution	Bivariate Log Normal Distribution
$\int_{-\infty}^{x_i} x\, h(x)\, dx$ (a)	$\int_{-\infty}^{x_i} w(x,y_k)\, dx$ (c)	$\int_{0}^{x_i} l(x)\, dx$ (i)	$\int_{0}^{x_i} z(x,y_k)\, dx$ (l)
$\int_{-\infty}^{x_i} x^2 h(x)\, dx$ (b)	$\int_{-\infty}^{x_i} x\, w(x,y_k)\, dx$ (d)	$\int_{0}^{x_i} x^{\alpha} l(x)\, dx$ (j)	$\int_{0}^{x_i} x\, z(x,y_k)\, dx$ (m)
	$\int_{-\infty}^{x_i} (x_i - x)\, w(x,y_k)\, dx$ (e)	$V(x)$ (k)	$\int_{0}^{x_i} \int_{0}^{y_j} x^{\alpha} z(x,y)\, dy\, dx$ (n)
	$\int_{-\infty}^{x_i} \int_{-\infty}^{y_j} x\, w(x,y)\, dy\, dx$ (f)		$\int_{0}^{x_i} \int_{0}^{y_j} x^{\alpha_1} y^{\alpha_2} z(x,y)\, dy\, dx$ (o)
	$\int_{-\infty}^{x_i} \int_{-\infty}^{y_j} xy\, w(x,y)\, dy\, dx$ (g)		
	$\int_{-\infty}^{u} \int_{-\infty}^{\infty} (u-v) g(v,u)\, dv\, du$ (h)		

Applying (B.6) and (B.8) to (B.14) and using (B.13) we have

$$\sigma^2 \int_{-\infty}^{v_i} v^2 f(v)\,dv + \mu^2\,F(v_i) - 2\mu\sigma\,f(v_i). \qquad (B.15)$$

The first term of (B.15) equals

$$\sigma^2 \int_{-\infty}^{v_i} -v\,df(v) = \sigma^2\,(F(v_i) - v_i\,f(v_i)). \qquad (B.16)$$

Substituting this result into (B.15) we find

$$(\mu^2 + \sigma^2)F(v_i) - \sigma\,(2\mu + \sigma v_i)f(v_i). \qquad (B.17)$$

c. $\displaystyle \int_{-\infty}^{x_i} w(x,y_k)\,dx = \int_{-\infty}^{v_i} \frac{1}{\sigma_y}\,g(v,u_k)\,dv$

$$= \frac{1}{\sigma_y} \int_{-\infty}^{v_i} \frac{1}{2\pi\sqrt{1-\rho^2}}\, e^{-\{((v-\rho u_k)/\sqrt{1-\rho^2})^2 + u_k^2\}/2}\,dv$$

$$= \frac{1}{\sigma_y}\,f(u_k)\,F(v_i^k). \qquad (B.18)$$

d. $\displaystyle \int_{-\infty}^{x_i} x\,w(x,y_k)\,dx = \int_{-\infty}^{v_i} \frac{(\sigma_x v + \mu_x)}{\sigma_y}\,g(v,u_k)\,dv$

$$= \frac{1}{\sigma_y}\,f(u_k) \int_{-\infty}^{v_i} \{\sigma_x(v - \rho u_k)$$

$$+ (\mu_x + \sigma_x \rho u_k)\}\frac{1}{\sqrt{2\pi(1-\rho^2)}}\, e^{-((v-\rho u_k)/\sqrt{1-\rho^2})^2/2}$$

$$= \{(\mu_x + \rho\sigma_x u_k)\,F(v_i^k)$$

$$- \sigma_x\sqrt{1-\rho^2}\,f(v_i^k)\}\,(f(u_k)/\sigma_y). \qquad (B.19)$$

e. $\displaystyle \int_{-\infty}^{x_i} (x_i - x)\,w(x,y_k)\,dx = \frac{\sqrt{1-\rho^2}\,\sigma_x\,f(u_k)}{\sigma_y}\,\{v_i^k\,F(v_i^k) + f(v_i^k)\}. \qquad (B.20)$

Calculating (B.20) the results (B.18) and (B.19) were used.

f. $\displaystyle \int_{-\infty}^{x_i} \int_{-\infty}^{y_j} x\,w(x,y)\,dy\,dx = \sigma_x \int_{-\infty}^{x_i} \int_{-\infty}^{y_j} \frac{(x - \mu_x)}{\sigma_x}\,w(x,y)\,dy\,dx$

$$+ \mu_x\,W(x_i,y_j). \qquad (B.21)$$

Applying transformations (B.6) and (B.7) and substitutions (B.8) and (B.9) we can write (B.21) as

$$\sigma_x \int_{-\infty}^{v_i} \int_{-\infty}^{u_j} v\, g(v,u)\, du\, dv + \mu_x\, G(v_i,u_j) = \sigma_x\, E_v^{v_i,u_j} + \mu_x\, G(v_i,u_j). \quad \text{(B.22)}$$

The term $E_v^{v_i,u_j}$ is calculated in the following way.

$$
\begin{aligned}
E_v^{v_i,u_j} &= -\int_{-\infty}^{u_j}\int_{-\infty}^{v_i} (\rho u - v) g(v,u)\, dv\, du + \rho\, E_u^{v_i,u_j}\\[4pt]
&= -\int_{-\infty}^{u_j}\int_{-\infty}^{v_i} (1-\rho^2)\frac{\partial}{\partial v}(g(v,u))\, du + \rho\, E_u^{v_i,u_j}\\[4pt]
&= -\int_{-\infty}^{u_j} \frac{(1-\rho^2)}{2\pi\sqrt{1-\rho^2}}\, e^{-((v_i^2 - 2\rho v_i u + u^2)/(1-\rho^2))/2}\, du + \rho\, E_u^{v_i,u_j}\\[4pt]
&= -(1-\rho^2)\, f(v_i)\, F(u_j^i) + \rho\, E_u^{v_i,u_j}. \quad \text{(B.23)}
\end{aligned}
$$

In the same way we obtain

$$E_u^{v_i,u_j} = -(1-\rho^2)\, f(u_j)\, F(v_i^j) + \rho\, E_v^{v_i,u_j}. \quad \text{(B.24)}$$

Combining (B.23) and (B.24) we have

$$
\begin{aligned}
E_v^{v_i,u_j} &= -\rho f(u_j)F(v_i^j) - f(v_i)F(u_j^i) \quad \text{and}\\
E_u^{v_i,u_j} &= -\rho f(v_i)F(u_j^i) - f(u_j)F(v_i^j). \quad \text{(B.25)}
\end{aligned}
$$

The final result thus equals

$$\mu_x\, G(v_i,u_j) - \sigma_x\, (\rho\, f(u_j)\, F(v_i^j) + f(v_i)\, F(u_j^i)). \quad \text{(B.26)}$$

g. $$\int_{-\infty}^{x_i}\int_{-\infty}^{y_j} xy\, w(x,y)\, dy\, dx = \sigma_x\, \sigma_y \int_{-\infty}^{x_i}\int_{-\infty}^{y_j} \frac{(x-\mu_x)(y-\mu_y)}{\sigma_x \sigma_y}\, w(x,y)\, dy\, dx$$
$$+ \mu_x\, E_y^{x_i,y_j} + \mu_y\, E_x^{x_i,y_j} - \mu_x \mu_y\, W(x_i,y_j). \quad \text{(B.27)}$$

Applying (B.6), (B.7), (B.8), (B.9) and using the result (B.26) we can write (B.27) as

$$\sigma_x \sigma_y \int_{-\infty}^{v_i}\int_{-\infty}^{u_j} vu\, g(v,u)\, du\, dv + \mu_x \mu_y\, G(v_i,u_j)$$
$$- \{f(v_i)F(u_j^i)(\rho\sigma_y\mu_x + \sigma_x\mu_y)$$
$$+ f(u_j)F(v_i^j)(\rho\sigma_x\mu_y + \sigma_y\mu_x)\}. \quad \text{(B.28)}$$

The first term of (B.28) equals

$$\int_{-\infty}^{v_i} \int_{-\infty}^{u_j} v u g(v,u) \, du \, dv$$

$$= \int_{-\infty}^{v_i} vf(v) \left[\int_{-\infty}^{u_j} \frac{u}{\sqrt{2\pi}\sqrt{1-\rho^2}} e^{-((u-\rho v)/\sqrt{1-\rho^2})^2/2} \, du \right] dv. \quad (B.29)$$

The term between the brackets can be written as

$$- \int_{-\infty}^{u_j} \frac{\sqrt{1-\rho^2}}{\sqrt{2\pi}} \frac{\partial}{\partial u} e^{-((u-\rho v)/\sqrt{1-\rho^2})^2/2} + \rho v \, F\left(\frac{u_j-\rho v}{\sqrt{1-\rho^2}}\right)$$

$$= -\sqrt{1-\rho^2} \, f\left(\frac{u_j-\rho v}{\sqrt{1-\rho^2}}\right) + \rho v \, F\left(\frac{u_j-\rho v}{\sqrt{1-\rho^2}}\right). \quad (B.30)$$

Multiplying (B.30) by $vf(v)$ and integrating over v the first term becomes

$$-\sqrt{1-\rho^2} \int_{-\infty}^{v_i} vf(v) \, f\left(\frac{u_j-\rho v}{\sqrt{1-\rho^2}}\right) dv$$

$$= -\sqrt{1-\rho^2} \int_{-\infty}^{v_i} \frac{v}{2\pi} e^{-((v-\rho u_j)/\sqrt{1-\rho^2})^2/2 - u_j^2/2} \, dv$$

$$= -\sqrt{1-\rho^2} \, f(u_j)(\rho u_j \sqrt{1-\rho^2} \, F(v_i^j) - (1-\rho^2)f(v_i^j)). \quad (B.31)$$

The second term becomes

$$\rho \int_{-\infty}^{v_i} v^2 f(v) \, F\left(\frac{u_j-\rho v}{\sqrt{1-\rho^2}}\right) dv$$

$$= -\rho \int_{-\infty}^{v_i} v F\left(\frac{u_j-\rho v}{\sqrt{1-\rho^2}}\right) df(v)$$

$$= -\rho v_i \, F(u_j^i) \, f(v_i) + \rho \int_{-\infty}^{v_i} f(v) F\left(\frac{u_j-\rho v}{\sqrt{1-\rho^2}}\right) dv$$

$$\quad - \frac{\rho^2}{\sqrt{1-\rho^2}} \int_{-\infty}^{v_i} vf(v) f\left(\frac{u_j-\rho v}{\sqrt{1-\rho^2}}\right) dv$$

$$= -\rho v_i \, F(u_j^i) \, f(v_i) + \rho \, G(v_i, u_j)$$

$$\quad - \frac{\rho^2}{\sqrt{1-\rho^2}} f(u_j)(\rho u_j \sqrt{1-\rho^2} \, F(v_i^j) - (1-\rho^2) \, f(v_i^j)). \quad (B.32)$$

Summing (B.31) and (B.32) we have

$$\int_{-\infty}^{v_i} \int_{-\infty}^{u_j} vu g(v,u) \, du \, dv = \rho \, G(v_i,u_j) - v_i \rho \, f(v_i) F(u_j^i)$$

$$- u_j \rho f(u_j) F(v_i^j) + (1 - \rho^2) \, g(v_i,u_j).$$

Finally, substitution of (B.33) into (B.28) gives   (B.33)

$$\int_{-\infty}^{x_i} \int_{-\infty}^{y_j} xy \, w(x,y) \, dy \, dx$$

$$= (\rho\sigma_x\sigma_y + \mu_x\mu_y) G(v_i,u_j) - (\sigma_x\sigma_y\rho v_i + \sigma_x\mu_y + \sigma_y\mu_x\rho) \, f(v_i) \, F(u_j^i)$$
$$- (\sigma_x\sigma_y\rho u_j + \sigma_y\mu_x + \sigma_x\mu_y\rho) f(u_j) F(v_i^j)$$
$$+ (1 - \rho^2) \, \sigma_x\sigma_y \, g(v_i,u_j). \tag{B.34}$$

h. $$\int_{-\infty}^{\infty} \int_{-\infty}^{u} (u - v) g(v,u) \, dv \, du. \tag{B.35}$$

Applying the transformations,

$$x = u - v \quad \text{and} \quad y = u + v \tag{B.36}$$

to (B.35) we obtain

$$\frac{1}{2} \int_{-\infty}^{\infty} \int_{0}^{\infty} \frac{x}{\sqrt{2\pi(1 - \rho^2)}} e^{-(1/2(1-\rho^2))(x^2(1+\rho)/2 + y^2(1-\rho)/2)} \, dx \, dy$$

$$= \frac{1}{2} \int_{-\infty}^{\infty} \frac{1}{\sqrt{2\pi(1 - \rho^2)}} e^{-y^2((1 - \rho)/(2 - 2\rho^2))/2}$$

$$dy \int_{0}^{\infty} \frac{x}{\sqrt{2\pi}} e^{-x^2((1+\rho)/(2-2\rho^2))/2} dx$$

$$= \frac{1}{2\sqrt{1 - \rho^2} \, (\sqrt{1 - \rho}/\sqrt{2 - 2\rho^2})} \cdot \frac{1}{(1 + \rho)/(2 - 2\rho^2)} \, f(0) = \frac{\sqrt{1 - \rho}}{\sqrt{\pi}} \, . \tag{B.37}$$

B.2 THE UNIVARIATE AND THE BIVARIATE LOG NORMAL DISTRIBUTION

B.2.1 Notation

The distribution,

$$l(x) = \frac{1}{\sigma\sqrt{2\pi}x} e^{-((\ln x - \mu)/\sigma)^2/2} \quad x > 0 \quad . \tag{B.38}$$

is called the log normal distribution (base e). The variable,

$$y = \ln x \tag{B.39}$$

has the normal distribution with mean μ and standard deviation σ. The bivariate log normal distribution is denoted by $z(x,y)$, where

$$z(x,y) = \frac{1}{2\pi\sigma_1\sigma_2\sqrt{1-\rho^2}\,xy}\exp\left[-\left(\left(\frac{\ln x - \mu_1}{\sigma_1}\right)^2\right.\right.$$
$$-2\rho\left(\frac{\ln x - \mu_1}{\sigma_1}\right)\left(\frac{\ln y - \mu_2}{\sigma_2}\right)$$
$$\left.\left.+\left(\frac{\ln y - \mu_2}{\sigma_2}\right)^2\right)/(2(1-\rho^2))\right]. \tag{B.40}$$

The variables $t = \ln x$ and $q = \ln y$ have the bivariate normal distribution with parameters $\mu_1, \mu_2, \sigma_1, \sigma_2$ and correlation coefficient ρ.

B.2.2 Transformations and Substitutions

We use the following transformations and substitutions in part B.2

$v = (\ln x - \mu_1)/\sigma_1,$	(B.41)	$u = (\ln y - \mu_2)/\sigma_2,$	(B.42)
$v_i = (\ln x_i - \mu_1)/\sigma_1,$	(B.43)	$u_j = (\ln y_j - \mu_2)/\sigma_2,$	(B.44)
$v_i^k = (v_i - \rho u_k)/\sqrt{1-\rho^2},$	(B.45)	$u_j^k = (u_j - \rho v_k)/\sqrt{1-\rho^2}.$	(B.46)

B.2.3 Some Derivations (See Table B.1)[2]

i.
$$\int_0^{x_i} l(x)\,dx.$$

Applying transformation (B.41) and substitution (B.43) we have

$$\int_0^{x_i} l(x)\,dx = F(v_i). \tag{B.47}$$

j.
$$\int_0^{x_i} x^\alpha l(x)\,dx = e^{\alpha\mu}\int_{-\infty}^{v_i} e^{\alpha\sigma v}\,f(v)\,dv$$

$$= e^{\alpha\mu}\int_{-\infty}^{v_i} e^{(\alpha\sigma)^2/2}\cdot\frac{e^{-(v-\alpha\sigma)^2/2}}{\sqrt{2\pi}}\,dv$$

$$= e^{\alpha\mu + (\alpha\sigma)^2/2}\,F(v_i - \alpha\sigma). \tag{B.48}$$

[2] μ equals μ_1 and σ equals σ_1.

k. $\quad V(x) = E(x^2) - E(x)^2 = e^{2\mu+2\sigma^2} - e^{2\mu+\sigma^2} = e^{2\mu+\sigma^2}(e^{\sigma^2}-1).$ (B.49)

l. $\qquad\qquad \int_0^{x_i} z(x,y_k)dx = \dfrac{1}{y_k\sigma_2} \int_{-\infty}^{v_i} g(v,u_k)dv.$ (B.50)

According to c), expression (B.50) equals

$$(1/y_k\sigma_2)\, f(u_k)\, F(v_i^k).$$ (B.51)

m. $\qquad \int_0^{x_i} x^\alpha z(x,y_k)dx$

$$= e^{\alpha\mu_1} \int_{-\infty}^{v_i} \frac{e^{\alpha\sigma_1 v}}{y_k\sigma_2} g(v,u_k)dv$$

$$= \frac{e^{\alpha\mu_1}}{y_k\sigma_2} f(u_k) \int_{-\infty}^{v_i} \frac{e^{\alpha\sigma_1 v}}{\sqrt{2\pi(1-\rho^2)}} e^{-((v-\rho u_k)/\sqrt{1-\rho^2})^2/2}\, dv.$$ (B.52)

Using the transformation, $t = (v - \rho u_k)/\sqrt{1-\rho^2}$ we can write (B.52) as

$$\frac{e^{\alpha\mu_1}}{y_k\sigma_2} f(u_k) \int_{-\infty}^{v_i^k} e^{\alpha\sigma_1(\sqrt{1-\rho^2}t+\rho u_k)} f(t)dt.$$ (B.53)

According to the result of j), expression (B.53) equals

$$e^{\alpha\mu_1 + \alpha\sigma_1\rho u_k + ((1-\rho^2)\alpha^2\sigma_1^2)/2} \frac{f(u_k)}{y_k\sigma_2} F(v_i^k - \sqrt{1-\rho^2}\alpha\sigma_1).$$ (B.54)

n. $\quad \int_0^{x_i}\int_0^{y_j} x^\alpha z(x,y)dy\,dx = e^{\alpha\mu_1} \int_{-\infty}^{v_i}\int_{-\infty}^{u_j} e^{\alpha\sigma_1 v} g(v,u)du\,dv.$ (B.55)

Since, $\alpha\sigma_1 v - (1/2(1-\rho^2))(v^2 - 2\rho uv + u^2)$ can be written as,

$$-\frac{1}{2(1-\rho^2)} \{(v-\alpha\sigma_1)^2 - 2\rho(v-\alpha\sigma_1)(u-\rho\alpha\sigma_1) + (u-\rho\alpha\sigma_1)^2\} + \frac{1}{2}\alpha^2\sigma_1^2$$

we may reduce (B.55) to

$$e^{\alpha\mu_1 + \alpha^2\sigma_1^2/2}\, G(v_i - \alpha\sigma_1, u_j - \rho\alpha\sigma_1).$$ (B.56)

o. $\qquad \int_0^{x_i}\int_0^{y_j} x^{\alpha_1}y^{\alpha_2} z(x,y)dy\,dx$

$$= e^{\alpha_1\mu_1 + \alpha_2\mu_2} \int_{-\infty}^{v_i}\int_{-\infty}^{u_j} e^{\alpha_1\sigma_1 v + \alpha_2\sigma_2 u} g(v,u)du\,dv.$$ (B.57)

The exponent behind the integral can be written as

$$
-\frac{1}{2(1-\rho^2)} \{(v - \rho\alpha_2\sigma_2 - \alpha_1\sigma_1)^2 - 2\rho(v - \rho\alpha_2\sigma_2 - \alpha_1\sigma_1)(u - \rho\alpha_1\sigma_1 - \alpha_2\sigma_2)
$$

$$
+ (u - \rho\alpha_1\sigma_1 - \alpha_2\sigma_2)^2\}
$$

$$
+ \frac{1}{2}\alpha_1^2\sigma_1^2 + \frac{1}{2}\alpha_2^2\sigma_2^2 + \rho\alpha_1\alpha_2\sigma_1\sigma_2.
$$

Hence, (B.57) reduces to

$$
e^{\alpha_1\mu_1 + \alpha_2\mu_2 + \alpha_1^2\sigma_1^2/2 + \alpha_2^2\sigma_2^2/2 + \rho\alpha_1\alpha_2\sigma_1\sigma_2} \, G(v_i - (\rho\alpha_2\sigma_2
$$

$$
+ \alpha_1\sigma_1), u_j - (\rho\alpha_1\sigma_1 + \alpha_2\sigma_2)). \tag{B.58}
$$

C CONTOUR LINES OF THE BIVARIATE LOG NORMAL DISTRIBUTION

The probability density function of the bivariate log normal distribution equals

$z(x,y) =$

$$\frac{1}{\sigma_1 \sigma_2 2\pi \sqrt{1 - \rho^2} xy} e^{\frac{-1}{2(1 - \rho^2)} \left\{ \left(\frac{\ln x - \mu_1}{\sigma_1} \right)^2 - 2\rho \left(\frac{\ln x - \mu_1}{\sigma_1} \right) \left(\frac{\ln y - \mu_2}{\sigma_2} \right) + \left(\frac{\ln y - \mu_2}{\sigma_2} \right)^2 \right\}}.$$

(C.1)

To determine the $\gamma\%$ contour line we must find the probability density function of z. We proceed as follows. The probability that $z(x,y)$ is larger than k, which we write as,

$$p(z(x,y) > k) \tag{C.2}$$

equals,

$$p \left[-\frac{1}{2}(v^2 - 2\rho vu + u^2)/(1 - \rho^2) > (\sigma_1 v + \sigma_2 u) + \ln (k\sigma_1 \sigma_2 2\pi \sqrt{1 - \rho^2}) + \mu_1 + \mu_2 \right] \tag{C.3}$$

where

$$v = (\ln x - \mu_1)/\sigma_1 \quad \text{and} \quad u = (\ln y - \mu_2)/\sigma_2. \tag{C.4}$$

230

As easily can be verified, (C.3) can be reduced to,

$$p[w^T A w < -2 \ln(k\sigma_1\sigma_2 2\pi\sqrt{1-\rho^2}) + \sigma_1^{\,2} + 2\rho\sigma_1\sigma_2$$
$$+ \sigma_2^{\,2} - 2(\mu_1 + \mu_2)] \tag{C.5}$$

where

$$w^T = [v + \sigma_1 + \rho\sigma_2, u + \sigma_2 + \rho\sigma_1] \quad \text{and} \quad A = \frac{1}{1-\rho^2}\begin{bmatrix} 1 & -\rho \\ -\rho & 1 \end{bmatrix}. \tag{C.6}$$

The random vector w has the multivariate normal distribution with covariance matrix

$$V = \begin{bmatrix} 1 & \rho \\ \rho & 1 \end{bmatrix}. \tag{C.7}$$

The necessary and sufficient condition for $w^T A w$ to have a chi square distribution is that

$$A V A = A. \tag{C.8}$$

Since A equals the inverse of V this condition is clearly satisfied. The non-centrality parameter of the chi square distribution equals,

$$E(w)^T A E(w) = \sigma_1^{\,2} + 2\rho\sigma_1\sigma_2 + \sigma_2^{\,2} \tag{C.9}$$

where $E(w)$ denotes the expectation of w. Furthermore, the number of degrees of freedom is two since the rank of A is two.

The cumulative distribution of $w^T A w$ can be computed with,

$$P(\chi^2 | \nu, \lambda) = e^{-\lambda/2} \sum_{j=0}^{\infty} \frac{(\lambda/2)^j}{j!} Q(\chi^2 | \nu + 2j) \tag{C.10}$$

where ν is the number of degrees of freedom and λ the noncentrality parameter. The distribution Q on the right-hand side of (C.10) is the cumulative central chi square distribution.

Using (C.10) we find for given β

$$P(w^T A w < \beta) = \gamma\%. \tag{C.11}$$

Putting the right-hand side of (C.5) equal to β, the value of k can be computed. This leads to

$$k = \frac{e^{-(\beta - (\sigma_1^{\,2} + 2\rho\sigma_1\sigma_2 + \sigma_2^{\,2}) + 2(\mu_1 + \mu_2))/2}}{\sigma_1\sigma_2 2\pi\sqrt{1-\rho^2}}. \tag{C.12}$$

Putting,

$$z(x,y) = k \qquad\qquad (C.13)$$

and solving (C.13) for x, we have the $\gamma\%$ contour line,

$$x = e^{\sigma_1 \{\rho u - \sigma_1(1 - \rho^2) \pm \sqrt{(1 - \rho^2)(\beta - (u + \rho\sigma_1 + \sigma_2)^2)}\}} + \mu_1$$

for

$$e^{\sigma_2(-\sqrt{\beta} - (\rho\sigma_1 + \sigma_2))} + \mu_2 \leqslant y \leqslant e^{\sigma_2(\sqrt{\beta} - (\rho\sigma_1 + \sigma_2))} + \mu_2. \qquad (C.14)$$

However, if one wishes to draw the contour line it is easier to change the variables to polar coordinates. This leads to

$$x = e^{\mu_1 + \sigma_1(\rho\sqrt{\beta}\cos(\zeta) + \sqrt{\beta(1 - \rho^2)}\sin(\zeta) - \sigma_1 - \rho\sigma_2)} \quad \text{and}$$

$$\text{for } 0 \leqslant \zeta \leqslant 2\pi$$

$$y = e^{\mu_2 + \sigma_2(\sqrt{\beta}\cos(\zeta) - \sigma_2 - \rho\sigma_1)} \qquad\qquad (C.15)$$

D A COMPUTER PROGRAM FOR THE LOCATION OF THE OPTIMAL SIZE PATTERN IF SHIFT RULE T IS BEING APPLIED

D.1 INTRODUCTION

In Chaps. 5, 6, 7 and 8 the recursion formulae were derived for a number of different combinations of probability density function and loss function. In most cases these formulae could only be solved by means of an iterative method. The writing of a computer program, which performs the iterations, will then usually be the easiest and fastest way to find the optimal size pattern. Such a program may be written with the purpose to solve one specific problem. It then will have a good performance but a limited applicability, because the loss function and the density function are fully integrated in the program and cannot easily be changed. To avoid this disadvantage, we have written a program which is able to receive, among others, the density function and the loss function as input. The conditions the loss function has to satisfy are equal to those mentioned in Chaps. 2 and 4. The program is written for standardization problems where shift rule T is to be applied. Only a few modifications, however, are necessary to extend its applicability to shift rule O. Because Simpson's rule is used to calculate the integrals, the c.p.u. time to find an optimal size pattern will be considerably larger than that of a program written for a specific problem. This reduces the usefulness of the program somewhat. It cannot for instance be used

233

to perform an extensive sensitivity analysis. The program is written in Basic and it was used on a 5100 I.B.M. desk computer. The program is, with an example of its output, listed in Sec. D.3. The input consists of the following parameters and functions.

1. Parameter values of the loss function and/or the probability density function.
2. The probability density function.
3. The function z_i.
4. The first derivative of the loss function V_2 with respect to x_i.
5. The first derivative of the loss function V_1 with respect to x_i.
6. The loss function V_2.
7. The loss function V_1.
8. The first derivative of z_i with respect to x_{i+1}.
9. The problem variables (see below).
10. The values of x_0 and x_{n+1}.
11. A starting set.

Except for the normal input after the "run" command the user has to (re)type and/or insert several statements. This requires some knowledge of Basic expressions.

D.2   DESCRIPTION OF THE INPUT

The subroutine beginning at statement (stmt) 1400 is called once by the main program. The subroutine can be used for the input of parameter values of the loss function and/or the probability density function. The loss function we used is defined by (2.7) and (2.8) and requires input of the parameters k_1, k_2, α (= A1) and β (= B1). Furthermore, the values of the constants that are used often in the program might be calculated here. In our case these are the values of $k_1^{1/\beta}$ (= K3) and $k_2^{1/\beta}$ (= K4) which repeatedly occur in the recursion formulae. The stmts 1450–1490 contain the print commands for the input parameters.

Next, the probability density function must be specified. This function is to be inserted at stmt 1520. The function value must be assigned to the variable F1. As can be seen, the density function used in the program listed in Sec. D.3 is the normal distribution with $\mu = 10$ and $\sigma = 1$.

The function z in the stmts 1550–1580 is the function z_i defined by (4.17). The (dummy) arguments X and Y denote respectively x_{i+1} and x_i. The constants K3 and K4 were calculated at 1430. In stmt 1590 the function FNR is defined. This function is a multipurpose function; it computes, dependent on the value of P1, respectively the function values of the functions mentioned above under point 4, 5, 6 and 7 and the probabilities of the service areas. The user must insert into 1620 and/or 1630 the first derivative of the loss function V_2 with respect to x_i. For the loss function (2.8) this expression becomes[1] $k_2(x_i^\alpha - x^\alpha)^{\beta-1}$. Note that in 1630 T3 is substituted for x_i and Y for x. If another loss function is specified the user must of course make the same substitution. In 1660 and 1670 we find the absolute value of the first derivative of the function V_1 with respect to x_i. Though the structural forms of V_1 and V_2 are equal to each other in our loss function, another V_1 can be used here. We could for instance take V_1 equal to $V_1 = k_1(e^x - e^{x_i})$ which would imply the stmt,

$$F3 = K1 * EXP(T3)$$

at 1660. Statement 1670 then has to be deleted.

The functions V_2 and V_1 must be inserted in respectively 1700 and 1730. Again the variables T3 and Y are used to denote x_i and x.

At 1790 the function FNO is defined. This function calculates the first derivative of z_i with respect to x_{i+1}. The arguments X and Y in this function denote respectively x_{i+1} and x_i. If the insertion of the functions mentioned above is completed the "run" command can be given. The program then asks the following information,

1. INPUT DENSITY FUNCTION

The user must type—between quotes—the name of the probability density function previously inserted at stmt 5010.

2. Input N, PRECISION, STEP, LAMBDA

N: The number of sizes.

PRECISION: The integrals are evaluated with the given relative precision. For most cases a value of 0.001 is sufficient.

STEP: This is the step size used for the calculation of the second derivatives. These are approximated by means of a formula of the form,

$$h''(y) = \frac{h'(y + STEP) - h'(y)}{STEP} .$$

Good results were obtained by taking STEP equal to 0.01 times the expectation of x.

LAMBDA: This is the variable λ described in Sec. 4.6.3. As a starting value one might take $\lambda = .01$.

---

[1] Assuming $x > 0$, we can omit the factor $\alpha\beta x_i^{\alpha-1}$ because it would appear elsewise on both sides of the equality sign in (4.15).

3. BOUNDS OF DENSITY FUNCTION

$X(0)$: x_0.
$X(N + 1)$: x_{n+1}.

If x_0 equals minus infinity and/or x_{n+1} equals plus infinity, it is recommended to choose $X(0)$ and $X(N + 1)$ such that about 99% of the demand is in the interval $(X(0), X(N + 1))$.

4. INPUT STARTING SET BY HAND (1 = YES)

If the answer to this question is any other number than one, the next question is,

4a. INPUT X(1), X(N)

After typing the starting values for x_1 and x_n the computer selects the remaining starting sizes $x_2, \ldots, x_{n-1}$ equally distributed over the interval (x_1, x_n).

If the answer to 4) is 1, the program continues with,

4b. INPUT STARTING SET

The user now has to type in the starting values of $x_1, \ldots, x_n$, separated by commas.

After this last input question the program starts with the iterations using the method described in Sec. 4.6.3. After every iteration the values of $E(L)$, λ and the sizes are printed.

If in two successive iterations, j and $j + 1$, the absolute value of

$$(E(L)^{j+1} - E(L)^j)/E(L)^{j+1}$$

is less than 10^{-6} the iteration process stops. The computer then finally calculates the probabilities of the service areas and prints these together with the values of z_i ($i = 2, \ldots, n$) and the vector of first derivatives. The values of this vector ought to be close to zero. If this is not the case the values of PRECISION and/or STEP have to be reduced.

D.3 Listing of Computer Program

```
0010 REM GENPUR / CALCULATES  OPTIMAL SIZE PATTERN
0020 REM ......./ USING SHIFT RULE T
0030 DIM Z(21),X(20,1),M(20,1),B(20,1),A(20,20),C(20,20),D(20)
0040 PRINT 'INPUT DENSITY FUNCTION '
0050 INPUT G$
0060 PRINT 'INPUT N,PRECISION,STEP,LAMBDA'
0070 INPUT N,E,K5,L
0080 MAT X(N,1)=(0)
0090 MAT M(N,1)=(0)
0100 MAT B(N,1)=(0)
0110 MAT A(N,N)=(0)
0120 MAT C(N,N)=(0)
0130 PRINT FLP,' PROBLEM VARIABLES'
0140 PRINT FLP,' N = ';N;'  PRECISION = ';E;'  STEP = ';K5;
0150 PRINT FLP,' LAMBDA = ';L
0160 PRINT 'INPUT X(0),X(N+1)'
0170 INPUT Z(1),Z(N+1)
0180 PRINT FLP,' '
0190 PRINT FLP,' PROBABILITY DENSITY FUNCTION IS ';G$
0200 PRINT FLP,' '
0210 PRINT FLP,' BOUNDS OF DENSITY FUNCTION'
0220 PRINT FLP,'  X(0)= ';Z(1);'  X(N+1) = ';Z(N+1)
0230 PRINT 'INPUT STARTING SET BY HAND (1=YES)'
0240 INPUT T1
0250 IF T1#1 GOTO 0290
0260 PRINT 'INPUT STARTING SET'
0270 MAT INPUT X
0280 GOTO 0350
0290 PRINT 'INPUT X(1),X(N)'
0300 INPUT X(1,1),X(N,1)
0310 S1=(X(N,1)-X(1,1))/(N-1)
0320 FOR I=2 TO N-1
0330 X(I,1)=X(I-1,1)+S1
0340 NEXT I
0350 GOSUB 1400
0360 PRINT FLP,'  E(L)   LAMBDA   SIZES'
0370 PRINT FLP,' '
0380 REM CALCULATE E(L)
0390 GOSUB 1030
0400 REM CALCULATE VECTOR OF FIRST DERIVATIVES
0410 E0=E1
0420 P2=0
0430 GOSUB 1200
0440 REM CALCULATE MATRIX OF SECOND DERIVATIVES
0450 A2=Z(1)
0460 FOR I=1 TO N
0470 B2=X(I,1)+K5
0480 T3=B2
0490 IF I=1 GOTO 0520
0500 T4=X(I-1,1)
0510 A2=FNZ(T3,T4)
0520 P1=0
0530 GOSUB 1840
0540 T5=T1
```

```
0550 A2=B2
0560 IF I≠N GOTO 0590
0570 B2=Z(N+1)
0580 GOTO 0670
0590 S1=Z(I+1)
0600 T3=X(I,1)
0610 T4=X(I+1,1)
0620 P1=1
0630 A(I,I+1)=-FNR(S1)*FNX(S1)*FNO(T4,T3)
0640 A(I+1,I)=A(I,I+1)
0650 T3=B2
0660 B2=FNZ(T4,T3)
0670 P1=1
0680 GOSUB 1840
0690 D(I)=SQR(ABS((T5-T1-M(I,1))/K5))
0700 M(I,1)=M(I,1)/D(I)
0710 A(I,I)=1
0720 NEXT I
0730 FOR I=1 TO N-1
0740 A(I,I+1)=A(I,I+1)/(D(I)*D(I+1))
0750 A(I+1,I)=A(I,I+1)
0760 NEXT I
0770 MAT C=A
0780 FOR I=1 TO N
0790 C(I,I)=C(I,I)+L
0800 NEXT I
0810 MAT C=INV(C)
0820 MAT B=C*M
0830 FOR I=1 TO N
0840 B(I,1)=B(I,1)/D(I)
0850 NEXT I
0860 MAT X=X-B
0870 REM CHECK POSITION SIZES
0880 IF X(1,1)<Z(1) GOTO 1000
0890 FOR I=2 TO N
0900 IF X(I,1)≤X(I-1,1) GOTO 1000
0910 NEXT I
0920 IF X(N,1)>Z(N+1) GOTO 1000
0930 REM CALCULATE E(L)
0940 GOSUB 1030
0950 L=L/10
0960 IF E1<0 GOTO 2180
0970 IF E1≥E0+1.E-11 GOTO 1000
0980 IF ABS((E1-E0)/E1)<1.E-06 GOTO 2200
0990 GOTO 0400
1000 L=L*100
1010 MAT X=X+B
1020 GOTO 0770
1030 REM SUBROUTINE FOR CALCULATION Z(I) (I=2,..N) AND E(L)
1040 S1=X(1,1)
1050 FOR I=2 TO N
1060 S2=X(I,1)
1070 Z(I)=FNZ(S2,S1)
1080 S1=S2
```

```
1090 NEXT I
1100 REM CALCULATE E(L)
1110 P2=2
1120 GOSUB 1200
1130 REM PRINT RESULTS
1140 PRINT USING FLP,2140,E1,L;
1150 FOR I=1 TO N
1160 PRINT USING FLP,2150,X(I,1);
1170 NEXT I
1180 PRINT FLP,' '
1190 RETURN
1200 REM SUBROUTINE FOR CALCULATION E(L) (P2=2) OR VECTOR OF
1210 REM FIRST DERIVATIVES (P2=0)
1220 E1=0
1230 FOR I=1 TO N
1240 B2=X(I,1)
1250 T3=B2
1260 A2=Z(I)
1270 P1=P2
1280 GOSUB 1840
1290 T4=T1
1300 A2=B2
1310 B2=Z(I+1)
1320 P1=P1+1
1330 GOSUB 1840
1340 IF P2=0 GOTO 1370
1350 E1=E1+T4+T1
1360 GOTO 1380
1370 M(I,1)=T4-T1
1380 NEXT I
1390 RETURN
1400 REM SUBROUTINE INPUT PARAMETERS LOSS FUNCTION
1410 PRINT 'INPUT K1,K2,ALPHA,BETA'
1420 INPUT K1,K2,A1,B1
1430 K3=K1↑(1/B1)
1440 K4=K2↑(1/B1)
1450 PRINT FLP,' '
1460 PRINT FLP,' PARAMETERS LOSS FUNCTION'
1470 PRINT FLP,'  K1 = ';K1;' K2 = ';K2;' ALPHA = ';A1;
1480 PRINT FLP,' BETA = ';B1
1490 PRINT FLP,' '
1500 RETURN
1510 DEF FNX(X)
1520 F1=.3989423*EXP(-(X-10)*(X-10)/2)
1530 RETURN F1
1540 FNEND
1550 DEF FNZ(X,Y)
1560 F2=((K4*X↑A1+K3*Y↑A1)/(K3+K4))↑(1/A1)
1570 RETURN F2
1580 FNEND
1590 DEF FNR(Y)
1600 GOTO 1650,1690,1720,1750 ON P1
1610 REM CALCULATE FIRST DERIVATIVE OF V2 WITH RESPECT TO X(I)
1620 F3=T3↑A1-Y↑A1
```

```
1630 F3=K2*F3↑(B1-1)
1640 RETURN F3
1650 REM CALCULATE FIRST DERIVATIVE OF V1 WITH RESPECT TO X(I)
1660 F3=Y↑A1-T3↑A1
1670 F3=K1*F3↑(B1-1)
1680 RETURN F3
1690 REM CALCULATE V2
1700 F3=K2*(T3↑A1-Y↑A1)↑B1
1710 RETURN F3
1720 REM CALCULATE V1
1730 F3=K1*(Y↑A1-T3↑A1)↑B1
1740 RETURN F3
1750 REM CALCULATE COVERAGES
1760 F3=1
1770 RETURN F3
1780 FNEND
1790 DEF FNO(X,Y)
1800 F4=(K4*X↑A1+K3*Y↑A1)/(K3+K4)
1810 F4=(F4↑(1/A1-1))*K4*(X↑(A1-1))/(K3+K4)
1820 RETURN F4
1830 FNEND
1840 REM SUBROUTINE NUMERICAL INTEGRATION / SIMPSON'S RULE
1850 S5=(B2-A2)/2
1860 A3=A2+S5
1870 T1=FNX(A2)
1880 T2=FNX(B2)
1890 S6=FNX(A3)
1900 S0=FNR(A2)*T1+FNR(B2)*T2
1910 S1=FNR(A3)*S6
1920 S2=0
1930 S6=1
1940 S7=S5*(S0+4*S1)/3
1950 S2=S2+S1
1960 S1=0
1970 S6=2*S6
1980 S8=S5
1990 S5=S5/2
2000 S9=A2-S5
2010 FOR I9=1 TO S6
2020 S9=S9+S8
2030 T2=FNX(S9)
2040 S1=S1+T2*FNR(S9)
2050 NEXT I9
2060 T1=S5*(S0+4*S1+2*S2)/3
2070 IF T1≠0 GOTO 2100
2080 T1=.0000001
2090 PRINT 'POSSIBLE ERROR IN NUMERICAL INTEGRATION, T1=0'
2100 IF ABS((T1-S7)/T1)≤E GOTO 2130
2110 S7=T1
2120 GOTO 1950
2130 RETURN
2140 :##.######  #||||
2150 :   -##.###
2160 :     -#||||
```

```
2170 :    #.####
2180 PRINT 'E(L)<0, ITERATION PROCESS STOPPED'
2190 STOP ERROR
2200 PRINT FLP,' '
2210 PRINT FLP,' ITERATION PROCESS ENDED'
2220 PRINT FLP,' '
2230 PRINT FLP,' COVERAGES'
2240 PRINT FLP,' '
2250 P1=4
2260 A2=Z(1)
2270 FOR I=2 TO N+1
2280 B2=Z(I)
2290 GOSUB 1840
2300 PRINT USING FLP,2170,T1;
2310 A2=B2
2320 NEXT I
2330 PRINT FLP,' '
2340 PRINT FLP,' '
2350 PRINT FLP,' VECTOR OF FIRST DERIVATIVES'
2360 PRINT FLP,' '
2370 FOR I=1 TO N
2380 PRINT USING FLP,2160,M(I,1);
2390 NEXT I
2400 STOP NORMAL
```

Example of Output

```
PROBLEM VARIABLES
  N =  2          PRECISION =  1E-3       STEP =  1E-2        LABDA =  1E-2

PROBABILITY DENSITY FUNCTION IS NORMAL (M=10)

BOUNDS OF DENSITY FUNCTION
  X(0)=  7.5    X(N+1) =  12.5

PARAMETERS LOSS FUNCTION
  K1 =  2       K2 =  1        ALPHA =  1      BETA =  1

   E(L)    LAMBDA     SIZES

  .703227 1E-02      9.000    11.000
  .869024 1E-02     10.621    11.400
  .740337 1E-01     10.350    11.263
  .604342 1E+00      9.587    10.984
  .603538 1E-01      9.639    10.999
  .603528 1E-02      9.645    11.004
  .603528 1E-03      9.646    11.004

ITERATION PROCESS ENDED

Z(I) I=2,...N

  10.098

COVERAGES

   .5329    .4546

VECTOR OF FIRST DERIVATIVES

  -5E-05   -2E-05
```

REFERENCES

Abramowitz, M. and I.A. Stegun, *Handbook of mathematical functions*, Dover, New York, 1972, pp. 257, 932–933, 940–941.

AKB, *Dikte van vlakglas voor gebouwen*, Adviesbureau voor Kwaliteitsbeleid en Besliskunde, Ir J.v. Ettinger- J. Sittig, Rotterdam, 1976.

Bard, Y., *Nonlinear parameter estimation*, Academic Press, New York, 1974.

Bobeszko, A., *Parameter gradation for machine elements and assemblies*, Second symposium on applications of mathematics to standardization, Warsaw, 1972.

Bruni, L., *International economies of scale with a given technique*, J. Indust. Statist., Vol XII, 1964, pp. 175–190.

Cohen, A.C., Jr., *Estimating the mean and variance of normal populations from singly truncated and doubly truncated samples*, Ann. Math. Statist., Vol. 21, 1950, pp. 557–569.

de Silva, L.L.S.S.K., *Standardization of flat steel cross sections*, Report on practical work with Bouwcentrum no. C. 546, Rotterdam, 1969.

Hausman, J.A. and D.A. Wise, *Social experimentation, truncated distributions, and efficient estimation*, Econometrica, Vol. 45, no. 4, 1977, pp. 919–938.

Hughes, E., *Electrical technology*, Longmans, Green and Co. Ltd., London, 1963, pp. 376–377.

Mey, J.L. and P.M.M.H. Snel, *Leerboek der bedrijfseconomie deel II*, Delwel, Den Haag, 1964.

242

Mood, A.M. and F.A. Graybill, *Introduction to the theory of statistics*, McGraw-Hill, New York, 1950, pp. 235–237.

OEEC, *Standardization, simplification and specialization in containers*, Organization for European Economic Co-operation, Paris, 1954.

Rozendal, P., *Optimale maatkeuze voor panties* (not published), Thesis, Erasmus University, Rotterdam, 1974.

Sanders, T.R.B., *The aims and principles of standardization*, International Organization for Standardization, Geneva, 1972, pp. 41–45.

S.B.R., *Voorkeur matenreeks voor (ter plaatse gestorte) betonkolommen*, Studiekommissie A 27, Stichting Bouwresearch, Rotterdam, 1979.

Sittig, J., *Defining quality costs*, Quality EOQC J., Vol. VII, no 3, 1963, pp. 67–74.

——, *Economically optimal standardization*, First symposium on applications of mathematics to standardization, Warsaw, 1969, pp. 22–34.

——, *Gradation of technical parameters in a standard*, Second symposium on applications of mathematics to standardization, Warsaw, 1972, pp. 175–209.

——, *Syllabus Operationeel Onderzoek*, Economische faculteit, Erasmus University, Rotterdam, 1973.

——, *Contribution of operations research to standardization*, Proceedings in O.R. 7, papers of the annual meeting D.G.O.R. 1977, Physica Verlag, Würzburg, Wien, 1978, pp. 326–335.

Verman, L.C., *Standardization; a new discipline*, Archon, Hamden, Connecticut, 1973, pp. 368–369.

Wagner, H.M., *Principles of operations research*, Prentice-Hall International, London, 1969, pp. 803–824.

Wouters, J., *Standaardisatie van de buitenafmetingen of hoofdafmetingen van houten raamkozijnen in de woningbouw*, Adviesbureau voor Kwaliteitsbeleid en Besliskunde, Ir J.v. Ettinger- J. Sittig, Rotterdam, 1969.

INDEX

244